Holt Mathematics

Chapter 11 Resource Book

HOLT, RINEHART AND WINSTON

A Harcourt Education Company

Orlando • **Austin** • New York • San Diego • London

ISBN 0-03-078308-9

6   170   09 08

CONTENTS

Holt Mathematics

Date ______________

Dear Family,

In this chapter, your child will learn about probability: how to work with experimental probability and theoretical probability and various applications of probability including independent/dependent events, combinations, and permutations. Probability will be related to genetics, ecology, and sports.

Probability is the measure of how likely an event is to occur. The more likely an event is to occur, the higher its probability. Experimental probability is one way of estimating the probability of an event. It is based on actual experiments or observations. **Experimental probability** is the ratio of the number of times an event occurs to the total number of trials or the number of times an experiment is carried out.

If Tanya made saves on 18 out of 25 shots, what is the experimental probability that she will make a save on the next shot?

$$P \approx \frac{\text{number of times an event occurs}}{\text{total number of trials}}$$

$$P(\text{save}) \approx \frac{\text{number of saves made}}{\text{number of shots attempted}}$$

$$\approx \frac{18}{25} \quad \textit{Substitute.}$$

The experimental probability that Tanya will make a save on the next shot is approximately $\frac{18}{25}$ or 72%.

The outcome of an experiment can be impossible, certain, or in between. If an event is impossible, it never happens and the probability is 0.

$$P(\text{impossible event}) = \frac{0}{\text{total number of trials}}$$

$$= 0 \quad \textit{0 divided by any number except 0 is}$$
$$\textit{equal to 0.}$$

If an event is certain, it always happens and the probability is 1.

Theoretical probability is used prior to an experiment or when it is not feasible to collect actual data.

Theoretical Probability

$$\text{probability} = \frac{\text{ways an event can occur}}{\text{total number of equally likely outcomes}}$$

Holt Mathematics

Find the probability of drawing the letter *E* from a bag of 100 Scrabble®
tiles. Write your answer as a fraction, a decimal, and a percent.

$$P = \frac{\text{number of ways an event can occur}}{\text{total number of equally likely outcomes}}$$

$P(E) = \dfrac{\text{number of } E\text{'s}}{\text{total number of tiles}}$ *Write the ratio.*

$= \dfrac{12}{100}$ *Substitute.*

$= \dfrac{3}{25}$ *Write in simplest form.*

$= 0.12 = 12\%$ *Write as a decimal and as a percent.*

The probability of drawing an *E* is $\frac{3}{25}$, 0.12, or 12%.

Your child will learn to find the probability of **independent** and
dependent events.

Erik rolls a 3 on one number cube and a 2 on another number cube.
Since the outcome of rolling one number cube does not affect the
outcome of rolling the second number cube, the events are
independent.

Tomoko chooses a 7th grader for his team from a group of 7th and
8th graders, and then Juan chooses a different 7th grader from the
remaining students. Since Juan can't pick the same student that
Tomoko picked, and since there are fewer students for Juan to
choose from after Tomoko chooses, the events are **dependent.**

Your child will also learn about **combinations** and **permutations**.

A **combination** is a grouping of items and events in which the order
does not matter. For example, if you want to find how many different
2-person teams can be made from a group of 4 people, you want to
find the number of combinations. One way to find the number of
combinations is to list each combination, and eliminate any
duplicates.

A **permutation** is an arrangement of objects or events in which the
order matters. For example, if you want to find how many different
numbers you can make with the digits 1, 2, 3, and 4, you want to
find the number of permutations. One way to find the number of
permutations is to use multiplication. To find the number of
permutations when there are 4 items, you multiply the following:

$4 \times 3 \times 2 \times 1 = 24.$

For additional resources, visit g.hrw.com and enter the keyword
MS7 Parent.

Holt Mathematics

LESSON
11-1 **Practice A**
Probability

Match each event to its likelihood.

1. rolling a number greater than 6 on a number cube labeled 1 through 6   **A** likely

2. flipping a coin and getting heads   **B** unlikely

3. drawing a red or blue marble from a bag of red marbles and blue marbles   **C** as likely as not

4. spinning a number less than 3 on a spinner with 8 equal sections marked 1 through 8   **D** impossible

5. rolling a number less than 6 on a number cube labeled 1 through 6   **E** certain

Solve.

6. A bag contains 4 red marbles, 3 green marbles, and 2 yellow marbles. The probability of randomly picking a yellow marble is $\frac{2}{9}$. What is the probability of not picking a yellow marble?   __________

7. A number cube is labeled 1 through 6. The probability of randomly rolling a 4 is $\frac{1}{6}$. What is the probability of not rolling a 4?   __________

Tell whether the event is impossible, unlikely, as likely as not, likely, or certain.

8. Janelle almost never eats meat. On Monday, the school cafeteria offers three main choices. The choices are hamburger, tuna, or a turkey sandwich. Estimate the probability that Janelle will choose a hamburger.   __________

9. Tyrone rides his bicycle to school if he gets up by 7:15 A.M. Tyrone gets up by 7:15 A.M. about half the time. Estimate the probability that Tyrone will ride his bicycle to school.   __________

3

Holt Mathematics

Practice B
Probability

Determine whether each event is impossible, unlikely, as likely as not, likely, or certain.

1. rolling an even number on a number cube labeled 1 through 6 _______________

2. picking a card with a vowel on it from a box of cards in which each letter of the alphabet is written on a card _______________

3. spinning a number greater than 2 on a spinner with 10 equal sections marked 1 through 10 _______________

4. drawing a red marble from a bag of black, blue, and green marbles _______________

5. flipping a coin and getting heads or tails _______________

6. rolling a number that is less than three 5 times in a row on number on a number cube labeled 1 through 6 _______________

Solve.

7. A bag contains 3 green marbles, 7 blue marbles, and 2 black marbles. The probability of randomly picking a green marble is $\frac{1}{4}$. What is the probability of not picking a green marble? _______________

8. A spinner has 8 equal sections labeled 1 through 8. The probability of spinning a number that is greater than or equal to 6 is $\frac{3}{8}$. What is the probability of spinning a number that is not greater than or equal to 6? _______________

9. The probability of randomly drawing a red card from a bag that contains red, blue, and green cards is $\frac{3}{10}$. What is the probability of not drawing a red card? _______________

10. Myra almost always spends at least 45 minutes on the treadmill. If Myra got on the treadmill at 5:20 P.M., estimate the probability that she will still be on the treadmill at 6:00. _______________

11. Morris rarely arrives home before 4:00 P.M. It is now 3:20 P.M. Estimate the probability that Morris will arrive home in the next 30 minutes. _______________

Holt Mathematics

Practice C
LESSON 11-1
Probability

Determine whether each event is impossible, unlikely, as likely as not, likely, or certain.

1. rolling a number less than 4 on a number cube labeled 1 through 6 _______________

2. picking a card with a multiple of 3 from a box with 10 number cards numbered 1 through 10 _______________

3. drawing a red marble from a bag of 3 blue marbles, 8 red marbles, and 2 green marbles _______________

4. rolling a number cube that has sides labeled 2, 4, 6, 8, 10, and 12, and getting an even number _______________

5. picking two cards that have a sum of greater than 10 from a set of number cards in which there are three 4's, five 3's, and two 5's _______________

6. drawing a yellow marble from a bag of 5 blue marbles, 2 green marbles, and 7 yellow marbles _______________

Solve.

7. A spinner has 16 sections labeled 1 through 16. The probability of spinning a number that is less than or equal to 10 is $\frac{5}{8}$. What is the probability of spinning a number that is not less than or equal to 10? _______________

8. The probability of randomly picking a striped marble from a bag that contains red, blue, and striped marbles marbles is $\frac{5}{12}$. What is the probability of not picking a striped marble? _______________

9. When Sam studies for 1 hour or more, he almost always gets at least a B on his math test. Sam studies from 8:45 to 10:00. Estimate the probability that Sam will get a B on his quiz. _______________

10. A bag contains 12 black cards and 11 red cards. Julia randomly draws 2 black cards and does not replace them. Will Julia be more likely to draw a black card than a red card on her next draw? Explain.

____

____

Holt Mathematics

LESSON 11-1 Reteach
Probability

You can describe the probability of an event as **impossible, unlikely, as likely as not, likely,** or **certain.**

For the spinner at the right:

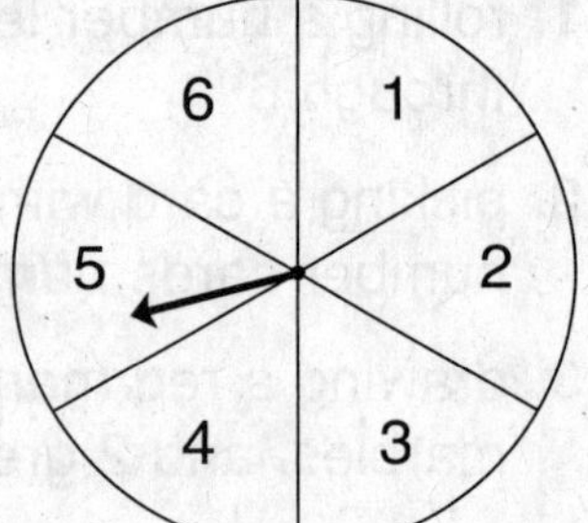

- Spinning a 7 is **impossible** because the spinner has no 7.
- Spinning a 5 is **unlikely** because only 1 of the 6 numbers is a 5.
- Spinning an even number is **as likely as not** because 3 of the numbers are even and 3 of the numbers are odd.
- Spinning a number that is greater than 1 is **likely** because 5 of the 6 numbers are greater than 1.
- Spinning a number that is less than 7 is **certain** because all of the numbers are less than 7.

Use the spinner. Write impossible, unlikely, as likely as not, likely, or certain to complete each statement.

1. Four of the 6 numbers are less than 5.

   Spinning a number that is less than 5 is _______________.

2. Six of the 6 numbers are greater than 0.

   Spinning a number that is greater than 0 is _______________.

3. One of the 6 numbers is a 3.

   Spinning a 3 is _______________.

A bag contains 1 red marble, 2 blue marbles, and 3 green marbles.

The probability of picking a red marble is $\frac{1}{6}$.
To find the probability of not picking a red marble,
subtract the probability of picking a red marble from 1. $P = 1 - \frac{1}{6} = \frac{5}{6}$

Solve.

4. A number cube is labeled 1 through 6. The probability of randomly rolling a 3 is $\frac{1}{6}$. What is the probability of not rolling a 3?

   $P = 1 - \frac{1}{6} =$ _______

5. A spinner has 5 sections labeled 1 through 5. The probability of randomly spinning an even number is $\frac{2}{5}$. What is the probability of not spinning an even number?

   $P = 1 -$ _______ $=$ _______

Holt Mathematics

Challenge
LESSON 11-1

A Likely Choice

There are 20 cards in a deck. One card is randomly selected. Provide additional information about the cards to describe each type of outcome.

Example: Describe a situation with a likely outcome.
There are 2 face cards and 18 number cards in the deck.
It is likely that you will select a number card.

1. Describe a situation with an unlikely outcome.

____

____

2. Describe a situation with a certain outcome.

____

____

____

3. Describe a situation with an outcome that is as likely as not.

____

____

4. Describe a situation with an impossible outcome.

____

____

5. A 1–6 number cube is rolled. Describe a roll with a likely outcome.

____

6. A 1–6 number cube is rolled. Describe a roll with an outcome that is as likely as not.

____

7. A 1–6 number cube is rolled. Describe a roll with an impossible outcome.

____

Holt Mathematics

<table>
<tr><td>LESSON
11-1</td><td></td></tr>
</table>

Problem Solving
Probability

Write the correct answer.

1. Of the original 56 signers of the Declaration of Independence, four of them represented North Carolina. If you selected one signer randomly, how likely is it that he represented North Carolina?

2. One question on a social studies multiple-choice test has four possible answers. Marianne is sure two of the choices are incorrect. How likely is she to choose the correct answer?

3. There are 8 right-handed pitchers and 2 left-handed pitchers on the Tigers baseball team. How likely is it that their opponents will face a right-handed pitcher?

4. Every seventh-grade student is attending a presentation about recycling in the auditorium. Jaleel is a seventh-grade student. How likely is it that he is in the auditorium?

Choose the letter for the best answer.

The table shows the contents of Leticia's CD and DVD collection.

Leticia's CD and DVD Collection

Type of CD/DVD	Number
Rock CD	26
Pop CD	17
Comedy DVD	11
Drama DVD	2

5. How likely is it that a disk chosen randomly is a CD?

  A unlikely   **C** certain

  B as likely as not   **D** likely

6. How likely is it that a disk chosen randomly is a jazz CD?

  F unlikely   **H** likely

  G as likely as not   **J** impossible

7. Sally picks a disk at random from Leticia's collection. How likely is it that it is a comedy DVD?

  A as likely as not

  B likely

  C certain

  D unlikely

8. Leticia picks three disks from her collection at random. Which outcome is impossible?

  F They are all rock CDs.

  G They are all drama DVDs.

  H None of them are rock CDs.

  J They are all DVDs.

Holt Mathematics

Name ___________________________________ Date ___________ Class ___________

Reading Strategies
Use Context

Probability measures how likely it is that an event will happen. Is it likely that a coin will land on heads when it is tossed?

An outcome is the result of an experiment. Landing on heads is a possible outcome of the experiment of tossing a coin.

Each outcome can be described in one of these ways:

• impossible   • unlikely   • as likely as not   • likely   • certain

Refer to the illustration for Exercises 1–5.

Complete each sentence. Write *impossible*, *unlikely*, *as likely as not*, *likely* or *certain*.

1. It is _______________ to draw a striped pebble from the bag.

2. You are _______________ to draw a pebble that is not black from the bag, because 9 of the 12 pebbles are not black.

3. If you reach into the bag, it is _______________ that you will draw a pebble.

4. Drawing a white pebble from the bag is _______________________, because 6 of the 12 pebbles are white.

5. Drawing a spotted pebble from the bag is _______________, because only 4 of the 12 pebbles are spotted.

Answer the following questions.

6. What word is used to measure the likelihood that an event will happen? _______________

7. What word means the result of an event? _______________

Holt Mathematics

Puzzles, Twisters & Teasers
LESSON 11-1 *Put Up Your Ducks!*

Circle words from the list in the word search (horizontally, vertically or diagonally). Find a word that answers the riddle and write it on the line.

trial measure probability outcome event
impossible likely unlikely certain experiment

```
I M P O S S I B L E P B E
N Q W E R T R I A L R O X
F A S D F G H J K L O U P
O Z X C V B N M I O B T E
R U N L I K E L Y U A C R
M E A S U R E V B N B O I
A P L M N K O I J B I M M
L I K E L Y F G H I L E E
B N C E R T A I N L I V N
U J I K O L E V E N T X T
A S D C H I C K E N Y G E
```

Why don't hens fight each other?

They're all ____________________________ .

10

Holt Mathematics

<table>
<tr><td>LESSON
11-2</td><td></td></tr>
</table>

Practice A
Experimental Probability

Find the experimental probability in the box. Each answer can be used only once.

$\frac{4}{11}$	$\frac{7}{9}$	$\frac{11}{15}$	$\frac{2}{9}$	$\frac{4}{15}$	$\frac{7}{11}$

1. Jolene is playing basketball. She scores on 11 out of the 15 baskets she shoots.

   a. What is the experimental probability that Jolene will get a basket on the next shot? ___________

   b. What is the experimental probability that Jolene will not get a basket on the next shot? ___________

2. Jamie is playing baseball. He gets a hit 7 out of 9 times at bat.

   a. What is the experimental probability that Jamie will get a hit his next time at bat? ___________

   b. What is the experimental probability that Jamie will not get a hit his next time at bat? ___________

3. Lou Ann is practicing for an archery tournament. She hits the target 7 out of 11 times.

   a. What is the experimental probability that Lou Ann will hit the target on the next shot? ___________

   b. What is the experimental probability that Lou Ann will not hit the target on the next shot? ___________

Find the experimental probability. Write your answer as a fraction, as a decimal, and as a percent.

4. A batter gets 6 hits in 12 times at bat. What is the experimental probability that she will get a hit in her next time at bat? ___________

5. A goalie blocks 16 out of 20 shots. What is the experimental probability that he will block the next shot? ___________

Holt Mathematics

Practice B
Experimental Probability

Find the experimental probability. Write your answer as a fraction, as a decimal, and as percent.

1. Jaclyn is a soccer goalie. If she has 21 out of 25 saves in practice, what is the experimental probability that she will have a save on the next shot on goal? _______________

2. If Harris hit the bull's-eye 3 out of 8 times at archery practice, what is the experimental probability that he will hit the bull's-eye on his next try? _______________

3. Nathan inspects new pants at a factory. Of the first 56 pairs of pants he inspected 49 were acceptable. What is the experimental probability that the next pair of pants will be acceptable? _______________

4. Sara has gone to work for 60 days. On 39 of those days she arrived at work before 8:30 A.M. On the rest of the days she arrived after 8:30 A.M. What is the experimental probability that she will arrive at work after 8:30 A.M. the next day she goes to work? _______________

Solve:

5. After a movie premiere, 99 of the first 130 people surveyed said they liked the movie.

   a. What is the experimental probability that the next person surveyed will say he or she liked the movie? _______________

   b. What is the experimental probability that the next person surveyed will say he or she did not like the movie? _______________

6. For the past 30 days, Naomi has been recording the number of customers at her restaurant between 10 A.M. and 11 A.M. During that hour, there have been fewer than 20 customers on 25 out of 30 days.

   a. What is the experimental probability that there will be fewer than 20 customers on the thirty-first day? _______________

   b. What is the experimental probability that there will be more than 20 customers on the thirty-first day? _______________

7. For the past four weeks, Nestor has been recording the daily high temperatures. During that time, the high temperature has been below 45° on 20 out of 28 days. What is the experimental probability that the high temperature will be below 45° on the twenty-ninth day? _______________

Holt Mathematics

Practice C
Experimental Probability

Find the experimental probability. Write your answer as a fraction, as a decimal, and as percent.

1. Luke is practicing his tennis serve. If he gets 21 out of 27 serves in, what is the experimental probability that he will get the next serve in?

   ____

2. Jose saw 50 people. Fourteen of them were wearing red shirts and 17 were wearing blue shirts. What is the experimental probability that the next person he sees will be wearing a blue shirt?

   ____

Solve.

3. During an exit survey after a play, 75 of the first 120 people surveyed said they did not like the play.

   a. What is the experimental probability that the next person surveyed will say he or she liked the play? _________________________

   b. What is the experimental probability that the next person surveyed will say he or she did not like the play? _________________________

4. For the past two weeks, Jimmy has been counting the number of joggers in the park between 8 P.M. and 9 P.M. each evening. In that time, there have been 40 or more joggers on 7 out of 14 days.

   a. What is the experimental probability that that there will be 40 or more joggers on the fifteenth day? _________________________

   b. What is the experimental probability that that there will be fewer than 40 joggers on the fifteenth day? _________________________

5. If Kathy hit the dartboard 9 out of 15 times and Toby hit the dartboard 14 out of 20 times, who has the greater experimental probability of hitting the dartboard on his or her next try?

   ____

6. Mona works at a snack bar. Of the first 25 hot dogs ordered one day, 19 were ordered with sauerkraut. If the snack bar expects to sell 150 hot dogs on any day, how many would they expect to be ordered with sauerkraut? _________________________

Holt Mathematics

<table><tr><td>LESSON
11-2</td><td></td></tr></table>

Reteach

Experimental Probability

Experimental probability is an estimate of the probability of an event. It is called *experimental* because you make observations or experiments to find the number of times a certain event actually happened.

$$\text{Experimental probability} \approx \frac{\text{number of times a certain event happened}}{\text{total number of trials}}$$

Suppose you have 12 marbles. Without replacing any marbles, you pull a red marble 5 times. What is the experimental probability of getting another red marble the next time you pull a marble?

$$P(\text{red}) \approx \frac{\text{number of red marbles}}{\text{number of marbles}} = \frac{5}{12}$$

The probability of getting red on the next pull is $\frac{5}{12}$.

At softball practice, Manny had 6 hits out of 15 times at bat.

1. What is the experimental probability that Manny will get a hit at his next time at bat?

$$P(\text{hit}) \approx \frac{\text{number of hits}}{\text{total number of times at bat}} = \frac{\quad}{\quad} = \frac{\quad}{\quad}$$

2. What is the experimental probability that Manny will not get a hit at his next time at bat?

$$P(\text{no hit}) \approx 1 - \frac{\text{number of hits}}{\text{total number of times at bat}} = 1 - \frac{\quad}{\quad} = \frac{\quad}{\quad}$$

Pam is playing darts. She hit the bull's eye 7 times out of 20 throws.

3. What is the experimental probability that Pam will hit the bull's eye on her next throw? ___________

4. What is the experimental probability that Pam will NOT hit the bull's eye on her next throw? ___________

So far this year Trisha's softball team has played 4 of their 20 games on Field A.

5. What is the experimental probability that they will play their next game on Field A? ___________

6. What is the experimental probability that they will play their next game on a field other than Field A? ___________

Holt Mathematics

<table><tr><td>LESSON
11-2</td><td>

Challenge
Pondering Probability
</td></tr></table>

Materials needed: paper plate,
construction paper, metal fastener

Use a paper plate, a construction paper
arrow, and a metal fastener to construct
a spinner. Your spinner should have 6 or
8 congruent sections. Draw a different
design in each section.

1. Based on the number of sections in your
   spinner, what fraction of the time would you predict that the
   pointer will stop on each section?

____

____

Spin the pointer 50 times and record
your results in the table below.

Section							
Times Spun							

2. What is your experimental probability of landing on each
   section?

____

____

3. Was your experimental probability of landing on any section
   close to your prediction? Explain.

____

____

4. Were any of your results surprising? Explain.

____

____

15

Holt Mathematics

Problem Solving
Experimental Probability

Write the correct answer as a fraction in simplest form.

This table shows a breakdown by format of total music sales in the United States in 2004.

Total American Music Sales in 2004

Format	Total (% of units shipped)
CD	80
Digital Single	15
Music Video	3
Other	2

1. What is the experimental probability that any random music purchase in 2004 was a CD?

2. What is the experimental probability that any random music purchase in 2004 was not a Music Video?

3. What is the experimental probability that any random music purchase in 2004 was a digital single?

4. Which combination of sales has an experimental probability of $\frac{1}{20}$?

Choose the letter for the best answer.

5. Ethan hits 4 ringers in 10 attempts while pitching horseshoes. What does an experimental probability of $\frac{2}{5}$ describe?

   A P(horseshoes)

   B P(missed shots)

   C P(attempts)

   D P(ringers)

6. Jay beats Terry at table tennis 3 out of 5 games. What is the experimental probability that Terry will win their next game?

   F $\frac{1}{2}$ H $\frac{2}{5}$

   G $\frac{3}{5}$ J 1

7. Poonam counts 10 classmates out of 36 people in the library. What is the experimental probability that the next person will be a classmate?

   A $\frac{5}{36}$ C $\frac{1}{36}$

   B $\frac{5}{18}$ D $\frac{1}{10}$

8. Macy makes 15 of 20 free throws at basketball practice. What is the experimental probability that she will miss her next free throw?

   F $\frac{1}{4}$ H $\frac{2}{3}$

   G $\frac{1}{2}$ J $\frac{3}{4}$

Holt Mathematics

 # Reading Strategies
Make Predictions

Experimental probability is a ratio. The ratio compares the number of times an event occurs to the total number of trials.

A trial is the number of times an experiment is carried out or an observation is made.

Experimental Probability

$$\text{probability} \approx \frac{\text{number of times an event occurs}}{\text{total number of trials}}$$

Refer to the illustration for Exercises 1 and 2. On this cube, you can land on 1, 2, or 3. Answer each question.

1. Predict which number you will land on most often. Explain.

____

2. Predict which number you will land on least often. Explain.

____

Actual events from an experiment may or may not match your prediction. The chart shows the outcomes from 100 trials.

Outcome	1	2	3
Toss	28	39	33

Refer to the table for Exercises 3 and 4. Answer each question.

3. Did your prediction for landing on 1 match the outcome in the table?

____

4. Did your prediction for landing on 3 match the outcome in the table?

____

Holt Mathematics

Puzzles, Twisters, & Teasers

LESSON 11-2 *All Bark, No Bite!*

Find each experimental probability. Write your answer as a fraction, a decimal, and a percent. Then match the letters to the answers to solve the riddle.

1. Nick hits a target 7 out of 20 times. What is the experimental probability that he will hit the target on his next try?

   N _________________

2. Melanie makes 12 out of 16 foul shots. What is the experimental probability that she will miss her next foul shot?

   B _________________

3. A pollster surveys 80 people to determine whether they will vote for Seth Gleason for mayor. 42 out of 80 say yes. What is the experimental probability that the next person surveyed will say he or she plans to vote for Seth Gleason? _________________

   K

4. A police radar measured the speed of 40 cars. The radar shows that 14 of the 40 cars are above the speed limit. What is the experimental probability that the next car will be traveling at or below the speed limit?

   A _________________

5. Rhonda watches students entering the school. She sees that 8 out of 25 are wearing boots. What is the experimental probability that the next student that Rhonda sees will be wearing boots?

   R _________________

6. A quarterback completes 10 of 16 passes. What is the experimental probability that his next pass will be incomplete? _________________

   I

Where does a dog keep its car?

IN A ____ ____ ____ ____ ____ ____ G LOT

0.25 65% 0.32 $\dfrac{21}{40}$ $\dfrac{3}{8}$ 35%

Holt Mathematics

Practice A
Make a List to Find Sample Spaces

1. Lindsay flips a coin and rolls a 1–6 number cube at the same time. What are the possible outcomes?

____

____

2. Jordan has a choice of wheat bread or rye bread and a choice of turkey, ham, or tuna for lunch. What are all the possible choices of sandwiches he can have?

____

____

3. Marisol has to decide whether to study Italian, French, or Spanish, and whether to take golf, tennis, or archery in gym class. What are the possible choices that Marisol has?

____

____

Choose the letter for the best answer.

4. Chad and Victoria are playing a game with a quarter and a spinner divided into sixths, numbered 1–6. Each player spins the spinner and tosses the coin. How many outcomes are possible in the game?

A 2   C 10

B 8   D 12

5. For a snack, Sophie can choose milk, apple juice, orange juice, or punch. To go with her drink, she can choose a chocolate cupcake, oatmeal cookie, or crackers. How large is the sample space?

F 12   H 4

G 7   J 3

6. Marva has a spinner divided into fourths and a 1–6 number cube. She spins the spinner and rolls the number cube. How many outcomes are possible in the game?

A 4   C 10

B 6   D 24

7. Larry has a choice of vanilla, chocolate, or strawberry ice cream. The choices of toppings are nuts, sprinkles, or coconut. How many one-topping sundaes can he make?

F 3   H 9

G 6   J 12

Holt Mathematics

Practice B
Make a List to Find Sample Spaces

1. Marcus spins the spinner at the right
   and flips a dime at the same time.
   What are the possible outcomes?
   How many outcomes are in the
   sample space?

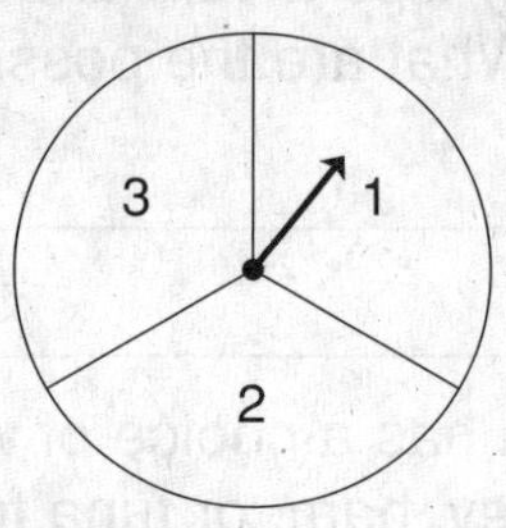

   ____

   ____

2. For lunch, Britney has a choice of a hot dog, a hamburger, or
   pizza and a choice of an apple, a pear, or grapes. What are all
   the possible choices of lunch she can have? How many
   outcomes are in the sample space?

   ____

   ____

   ____

3. Susan and Ryan are playing a game
   that involves spinning the spinner at
   the right and flipping a penny. How
   many outcomes are possible in the
   game?

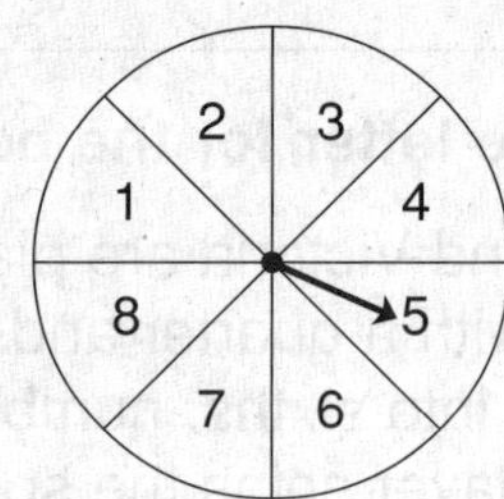

   ____

4. An Italian restaurant offers small, medium, and large calzones.
   The choices of fillings are cheese, sausage, spinach, or
   vegetable. How many different calzones can you order?

   ____

5. There are 5 ways to go from Town X to Town Y. There are 3
   ways to go from Town Y to Town Z. How many different ways
   are there to go from Town X to Town Z, passing through Town
   Y?

   ____

6. Rasheed has tan pants, black pants, gray pants, and blue pants.
   He has a brown sweater and a white sweater. How many
   different ways can he wear a sweater and pants together?

   ____

Holt Mathematics

Practice C
Make a List to Find Sample Spaces

1. Joanna spins the spinners at the right at the same time. What are the possible outcomes? How many outcomes are in the sample space?

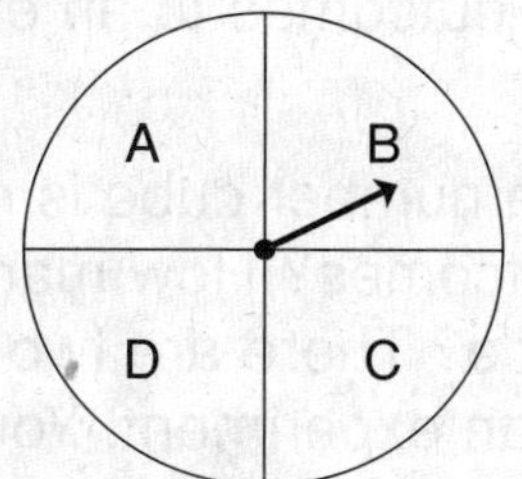

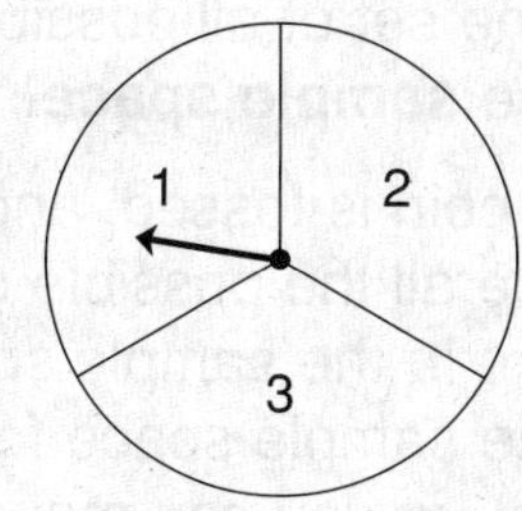

____

____

____

2. For breakfast, Armando has a choice of pancakes, eggs, or cereal and a choice of milk, hot cocoa, or juice. What are all the possible choices of breakfast he can have? How many outcomes are in the sample space?

____

____

3. Shannon and Tyler are playing a game that involves spinning the spinner shown at the right and tossing a 1–6 number cube. How many outcomes are possible in the game?

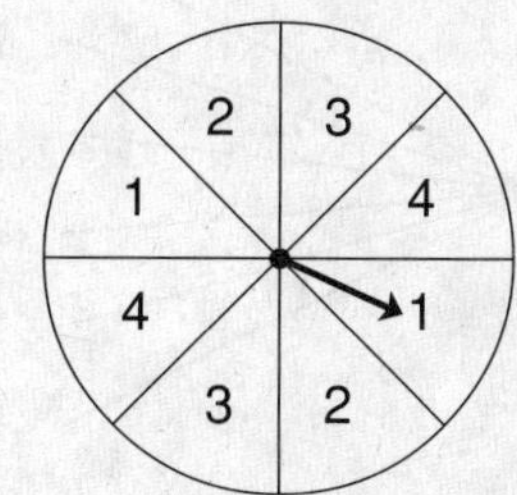

____

4. If you flip a penny, toss a 1–6 number cube, and flip a quarter, how many outcomes are possible? ___________________________

5. A Chinese restaurant has a special on Friday nights. For $20, you can choose one dish from 6 choices in column A and one dish from 5 choices in Column B. In addition, you can choose egg drop or wonton soup. How many different specials can you order? ___________________________

6. Lisa has a beige skirt, a black skirt, and a denim skirt. She has a red sweater and a white sweater, and she has a white blouse, a blue blouse, and a green blouse. How many different ways can she wear a skirt, sweater, and blouse together? ___________________________

Holt Mathematics

LESSON 11-3 Reteach
Make a List to Find Sample Spaces

The set of all possible outcomes to an experiment is called
the **sample space**.

A coin is tossed, and a number cube is rolled. What
are all the possible outcomes? How many outcomes
are in the sample space? There are two ways to show
the sample space for an experiment. You can make a
list, or you can make a tree diagram.

Make a list.

1. The possible outcomes for tossing a coin are _________ (H) and _________ (T).

2. The possible outcomes for rolling a number cube are ____, ____, ____, ____,

   ____, and ____

3. The sample space is (H, 1), (H, 2), (H, 3), (H, 4), (H, 5), (H, 6), (T, 1), (T, 2),

   (T, 3), (T, 4), (T, 5), (T, 6). There are ____ possible outcomes in the sample space.

Make a tree diagram.

4.

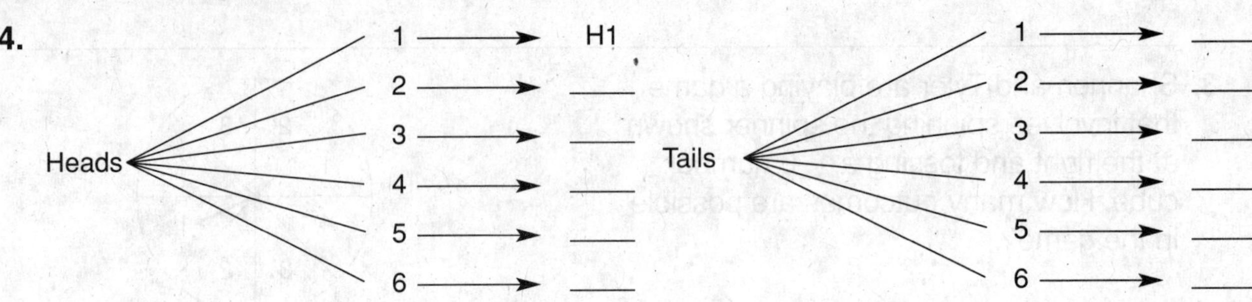

You can also find the number of possible outcomes by using the
Fundamental Counting Principle.

Multiply the possible outcomes of each event.

5. flipping the coin   rolling the number cube

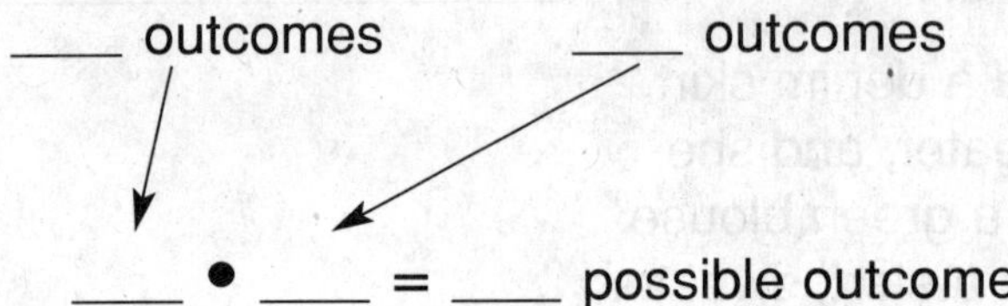

   ____ outcomes   ____ outcomes

   ____ • ____ = ____ possible outcomes

Holt Mathematics

LESSON
11-3

Challenge
Mutually Exclusive Cards?

Two events are *mutually exclusive* if they cannot occur at the same time.

Example: In a deck of 52 cards, there are 26 red and 26 black cards.

> *Event A*: Draw a red card. *Event B*: Draw a black card. The events are mutually exclusive. They cannot occur at the same time because a card cannot be red and black.

> *Event A*: Draw a red card. *Event B*: Draw a number card. The events are not mutually exclusive. They can occur at the same time since a number card may be red.

For each pair of events, tell whether the two events are mutually exclusive for a single experiment. If they are not, explain why.

1. In a deck of cards, there are 40 number cards and 12 face cards. *Event A*: Draw a number card. *Event B*: Draw a face card.

   ____

2. In a deck of cards, there are 26 black cards and 12 face cards. *Event A*: Draw a black card. *Event B*: Draw a face card.

   ____

3. In a deck of cards, there are 4 kings and 4 queens. *Event A*: Draw a king. *Event B*: draw a queen.

   ____

4. In a deck of cards, there are 12 diamonds and 12 hearts. *Event A*: Draw a diamond. *Event B*: Draw a heart.

   ____

5. In a deck of cards, there are 12 diamonds and 12 face cards. *Event A*: Draw a diamond. *Event B*: Draw a face card.

   ____

6. In a deck of cards, there are 40 number cards and 4 jacks. *Event A*: Draw a number card. *Event B*: Draw a jack.

   ____

7. In a deck of cards, there are 4 tens and 13 clubs. *Event A*: Draw a ten. *Event B*: Draw a club.

   ____

Holt Mathematics

Problem Solving
Make a List to Find Sample Spaces

Write the correct answer.

1. If you order one topping, how many different choices of bagel and toppings can you order?

____

2. Santana only likes cream cheese or jam on his bagel. How many choices does he have for a one-topping bagel?

____

3. Yesterday, Benny ran out of raisin bagels. How many choices of a bagel and one topping were there?

____

4. Today, Benny has all 5 types of bagels but runs out of honey. How many choices of a bagel with one topping can you order?

____

Benny's Bagels

Bagels	Toppings
Plain	Cream cheese
Poppy	Honey
Raisin	Butter
Sesame	Jam
Egg	

Choose the letter for the best answer.

5. The mall movie multiplex is showing 12 movies. Each movie is shown at five different times during the day. How many choices of movies and showtimes does Reggie have?

A 5　　　　　**C** 17

B 12　　　　**D** 60

6. At Hi-Top Ski Resort, there are three chair lifts to the top of the mountain. There are six ski trails to the bottom of the mountain. How many possible choices of lifts and trails do the skiers have?

F 9　　　　　**H** 81

G 18　　　　**J** 2

7. In a Little League game, Geri can bat first, second, or third. When at bat, she could strike out, walk, or get a hit. How many outcomes are in the sample space for these events?

A 3

B 6

C 9

D 18

8. Ty is flipping a coin. He has decided that if he flips the same result twice in a row, he will do his homework. If he flips 2 different results, then he will go jogging. How likely is it that he will study?

F as likely as not

G likely

H unlikely

J certain

Holt Mathematics

Reading Strategies
Read a Chart

To measure the probability of an outcome, you must find all the possible outcomes to the experiment. A **sample space** lists all possible outcomes to an experiment.

This spinner can land on white or black.

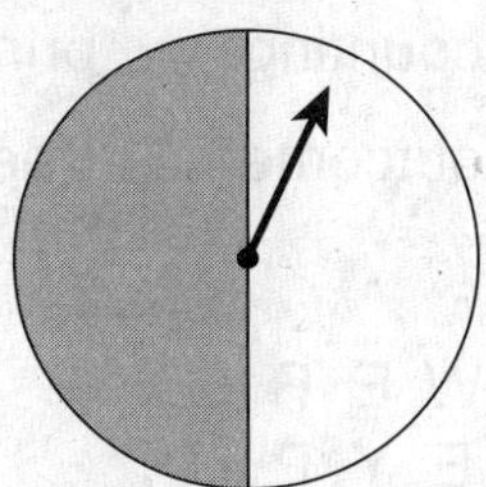

Each line of this chart lists a possible outcome from two spins.

Spin 1	Spin 2
Black	Black
White	Black
Black	White
White	White

Use the chart for Exercises 1–5. Answer each question.

1. What does a sample space help do?

____

2. With an outcome of black on Spin 1, what outcomes are possible on Spin 2?

____

3. With an outcome of black on Spin 1, one outcome for Spin 2 is black. This is shown as "black–black." How would you write the other possible outcome after Spin 2?

____

4. With an outcome of white on Spin 1, what are the possible outcomes for Spin 2?

____

5. How many possible outcomes are there for two spins?

____

Holt Mathematics

<table>
<tr><td>LESSON
11-3</td><td></td></tr>
</table>

Puzzles, Twisters & Teasers
Sealed With a Fish

Circle words from the list in the word search (horizontally, vertically or diagonally). Find a word that answers the riddle and write it on the line.

sample	space	counting	principle	fundamental
determine	possible	outcome	tree	diagram

```
D E T E R M I N E W E R O
I P R I N C I P L E Y P U
A Q N Z T C V B N M K L T
G W E W E R T Y U I O S C
R A S D F G E J K L P E O
A Q A Z X S W E D C V A M
M F U N D A M E N T A L E
N Z X C V B G T S H U J I
C O U N T I N G I P O L T
R C V B G T Y H N J A I K
P O S S I B L E N J U C D
S A M P L E V G T E D F E
```

What is gray, eats fish, and lives in Washington, D.C.?

The presidential ___ ___ ___ ___ .

Holt Mathematics

Practice A
Theoretical Probability

Tina has 3 quarters, 1 dime, and 6 nickels in her pocket. Find the probability of randomly drawing each of the following coins. Write your answer as a fraction, as a decimal, and as a percent.

		Fraction	Decimal	Percent
1.	quarter			
2.	dime			
3.	nickel			

Find the probability of each event. Write your answer as a fraction, as a decimal, and as a percent. Round to the nearest tenth of a percent.

4. randomly choosing a red card in a game that has 10 red, 10 blue, 10 green, 10 yellow cards, and 10 orange cards

____

5. tossing two fair coins and having both land tails up

____

6. randomly drawing 1 of the 4 S's from a bag of 100 Scrabble tiles

____

7. rolling a number greater than 4 on a fair number cube

____

A game has 12 blue disks, 10 red disks, and 8 black disks. Find the probability of each event when a disk is chosen at random.

8. red _________________   9. black _________________

10. blue _________________   11. not red or blue _________________

Holt Mathematics

Practice B
Theoretical Probability

**Find the probability of each event. Write your answer as a
fraction, as a decimal, and as a percent. Round to the nearest
tenth of a percent.**

1. randomly choosing a white counter from a bag of 12 red
   counters, 12 white counters, 12 green counters, and 12 blue
   counters

   ____

2. tossing two fair coins and having one land on tails and one land
   on heads

   ____

3. rolling a number greater than 1 on a fair number cube

   ____

4. randomly drawing an orange disk from a bag of 14 black disks,
   4 blue disks and 12 orange disks

   ____

5. randomly drawing 1 of the 6 R's from a bag of 100 Scrabble tiles

   ____

6. spinning a number less than 7 on a fair spinner with 8 equal
   sections labeled 1-8

   ____

**A set of cards has 20 cards with stars, 10 cards with squares,
and 15 cards with circles. Find the probability of each event
when a card is chosen at random.**

7. square ___________   8. circle ___________

9. star or circle ___________   10. not circle or square ___________

**There are 14 girls and 18 boys in Ms. Wiley's class. Ms. Wiley
randomly selects one student to solve a problem. Find the
probability of each event.**

11. selecting a boy ___________   12. selecting a girl ___________

Holt Mathematics

<table><tr><td>LESSON
11-4</td><td><h1>Practice C</h1>
Theoretical Probability</td></tr></table>

Find the probability of each event. Write your answer as a fraction, as a decimal, and as a percent. Round to the nearest tenth of a percent.

1. tossing three fair quarters and having all three of them land on heads

   ____

2. randomly choosing a classical CD from a collection of CDs consisting of 35 jazz CDs, 20 classical CDs, 25 rock CDs, and 5 country music CDs

   ____

3. randomly choosing a card with an even number from a shuffled deck of 52 cards with four 13-card suits (diamonds, hearts, clubs and spades), each of which has 9 number cards labeled 2-10 and 4 other cards

   ____

4. randomly drawing a vowel from a bag of 100 Scrabble® tiles that has 12 E's, 9 I's, 8 O's, 4 U's and 2 Y's.

   ____

There are 15 girls and 9 boys in Anne's yoga class. One of them is randomly selected to demonstrate a yoga position. Find the probability of each event.

5. selecting a boy ___________

6. selecting a girl ___________

Find the probability of each event when two 1-6 number cubes are rolled.

7. P(total of 5) ___________

8. P(total of 10) ___________

9. P(total ≥ 7) ___________

10. P(total < 2) ___________

Holt Mathematics

LESSON 11-4 Reteach
Theoretical Probability

The **theoretical probability** of an event is found by comparing the number of ways an event can occur to the total number of equally likely outcomes.

$$\text{theoretical probability} = \frac{\text{number of ways the event can occur}}{\text{total number of equally likely outcomes}}$$

One of the games at a carnival is the Wheel of Letters. Find the probability that the wheel will stop on each letter. Write your answer as a fraction, as a decimal, and as a percent.

1. The spinner has ______ equal sections. Each section is an equally likely outcome.

2. There is ______ section marked A.

3. There are ______ sections marked B.

4. There are ______ sections marked C.

5. There are ______ sections marked D.

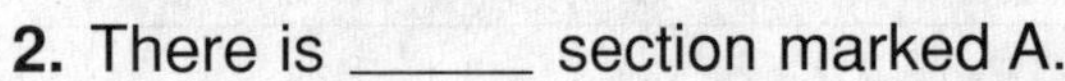

6. $P(B) =$ ___ $= 0.3 =$ ________ %

7. $P(A) =$ ___ $= 0.1 =$ ________ %

8. $P(C) =$ ___ $=$ ___ $= 0.4 =$ ________ %

9. $P(D) =$ ___ $=$ ___ $= 0.2 =$ ________ %

There are 11 pennies and 9 dimes in a bag. Find the probability of each event. Write your answer as a fraction, as a decimal, and as a percent.

10. Find the probability that a dime will be drawn from the bag.

$P(\text{dime}) =$ ——— $= 0.45 =$ ________ %

11. P(penny) = ________ $= 0.55 =$ ________

There are 6 yellow cards, 4 blue cards, and 10 green cards in a bag. A card is chosen at random. Find the probability of each event.

12. yellow ________

13. blue ________

14. green ________

15. blue or green ________

Holt Mathematics

<table>
<tr><td>LESSON
11-4</td><td>Challenge
Spinner Sums</td></tr>
</table>

Suppose these three spinners are each spun once.

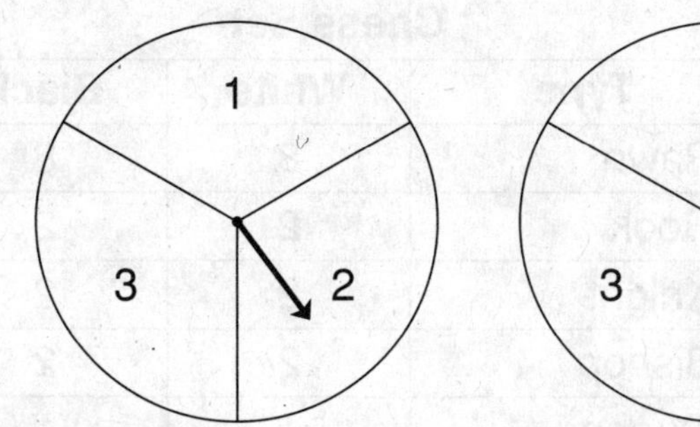
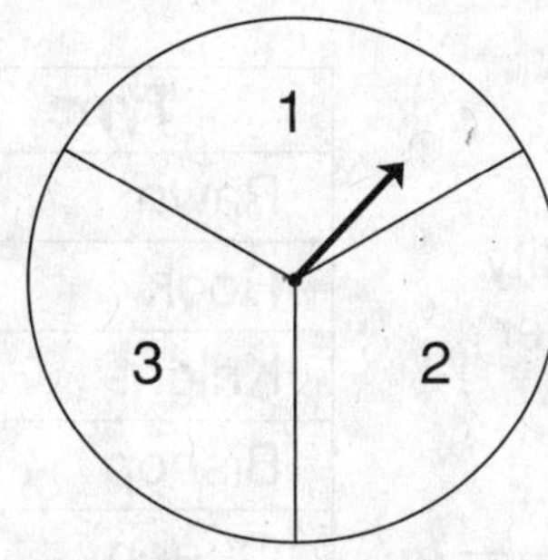
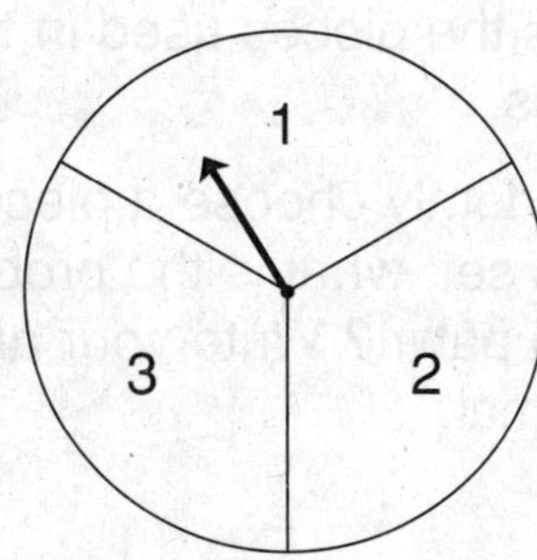

1. Complete the organized list so that it shows every sum that is possible if all three spinners are spun once.

First Spinner Shows 1	$1 + 1 + 1 = 3$ $1 + 1 + 2 = 4$ $1 + 1 + 3 = 5$	$1 + 2 + 1 = 4$ $1 + 2 + 2 = 5$ $1 + 2 + 3 = 6$	$1 + 3 + 1 = 5$ $1 + 3 + 2 = 6$ $1 + 3 + 3 = 7$
First Spinner Shows 2	$2 + 1 + 1 = 4$ ________ ________	________ ________ ________	________ ________ ________
First Spinner Shows 3	________ ________ ________	________ ________ ________	________ ________ ________

2. Complete the table using the chart of sums in Exercise 1.

Sum of Spinners	3	4	5	6	7	8	9
Number of Ways to Get the Sum							

3. How many possible outcomes are there are if you spin the three spinners? ____________

Using the table in Exercise 2, give the probability of getting each sum. Write the probability to the nearest tenth of a percent.

4. $P(\text{sum of 5})$ **5.** $P(\text{sum} > 7)$ **6.** $P(\text{sum} < 8)$

________________ ________________ ________________

Holt Mathematics

Problem Solving
Theoretical Probability

Write the correct answer in simplest form.

The table lists the pieces used in the game of chess.

Chess Set

Type	White	Black
Pawn	8	8
Rook	2	2
Knight	2	2
Bishop	2	2
Queen	1	1
King	1	1

1. If you randomly choose a piece of the chess set, what is the probability that it is a pawn? Write your answer as a decimal.

2. If you randomly choose a piece of the chess set, what is the probability that it is a white pawn? Write your answer as a decimal.

3. If you randomly choose a piece of the chess set, what is the probability that it is a rook, knight, or bishop? Write your answer as a fraction.

4. If you randomly choose a piece of the chess set, what is the probability that it is a queen? Write your answer as a fraction.

Choose the letter for the best answer.

5. Mr. Rose draws names to see who will give the first book report. There are 10 boys and 14 girls in his class. What is the probability that he will draw a girl's name?

   A $\frac{2}{5}$ C $\frac{7}{12}$

   B $\frac{5}{12}$ D $\frac{5}{7}$

6. Eight students will give reports on novels, 9 will report on biographies, and 7 will report on history books. What is the probability that the first report will be a novel?

   F $\frac{3}{8}$ H $\frac{8}{17}$

   G $\frac{1}{3}$ J $\frac{1}{2}$

7. Stanley is reading a 224-page book. There are illustrations on 14 pages. If Stanley opens the book at random, what is the probability that the page will have an illustration?

   A 0.0004 C 0.0714

   B 0.0625 D 0.9375

8. In Congress, each of the 50 states is represented by 2 senators. If you choose a senator randomly, what is the probability that you will choose a senator that represents Virginia?

   F 25% H 2%

   G 10% J 1%

Holt Mathematics

Reading Strategies
Interpret Information

When you are conducting an experiment, a **favorable outcome** is an outcome you are looking for.

Theoretical Probability

$$\text{probability(event)} = \frac{\text{number of ways an outcome can occur}}{\text{total number of possible outcomes}}$$

When you toss a number cube, there are 6 possible outcomes. There is only one way to get a 4 on the cube, so there is one favorable outcome for rolling a 4.

Read: The probability of rolling a 4 is 1 out of 6.

Write: $P(\text{rolling a 4}) = \frac{1}{6}$

Answer each question.

1. What is the probability of rolling a two on the cube?

   ____

2. What are the even numbers on the cube?

   ____

3. What is the probability of rolling an even number?

   ____

4. What are the odd numbers on the cube?

   ____

5. What is the probability of rolling an odd number.

   ____

If events have the same chance of happening, they are **equally likely** to occur.

6. Is the probability of rolling a 5 or an even number equally likely?

   ____

7. Is the probability of rolling an odd or an even number equally likely?

   ____

Holt Mathematics

Puzzles, Twisters & Teasers
Probably Problems!

Across

1. _____ probability is used to calculate the probability of an event when all outcomes are equally likely.

3. You can write a _____ as a decimal or a fraction.

5. If each possible outcome of an experiment is equally likely, the experiment is said to be _____.

7. You can write a _____ as a fraction or a percent.

8. A _____ outcome is one that you are looking for when you conduct an experiment.

Down

2. _____ probability compares the number of times an event occurs to the total number of trials.

4. You can write _____ as a fraction, a decimal, or a percent.

5. You can write a _____ as a decimal or as a percent.

Holt Mathematics

Practice A
Probability of Independent and Dependent Events

**Decide if each set of events is independent or dependent.
Explain your answer.**

1. A student spins a spinner and rolls a number cube.

   ____

   ____

2. A student picks a raffle ticket from a box and then picks a
   second raffle ticket without replacing the first raffle ticket.

   ____

   ____

**Find the probability of each set of independent events. Choose
the letter for the best answer.**

3. drawing a black checker from a bag
   of 6 black checkers and 4 red
   checkers, replacing it, and drawing
   another black checker

   A $\frac{2}{3}$ **C** $\frac{2}{5}$

   B $\frac{9}{25}$ **D** $\frac{3}{5}$

4. rolling a six on the first roll of a 1–6
   number cube and rolling an odd
   number on the second roll of the
   same cube

   F $\frac{1}{12}$ **H** $\frac{1}{6}$

   G $\frac{1}{8}$ **J** $\frac{1}{2}$

5. flipping a tail on a coin and spinning
   a 5 on a spinner with sections of
   equal area numbered 1–5

   A $\frac{1}{2}$ **C** $\frac{1}{7}$

   B $\frac{1}{5}$ **D** $\frac{1}{10}$

6. drawing a 1, 2, or 3 from 9 cards
   numbered 1–9, replacing the card,
   and drawing a 7, 8, or 9

   F $\frac{1}{3}$ **H** $\frac{1}{9}$

   G $\frac{3}{8}$ **J** $\frac{1}{12}$

Solve.

7. There are 4 black marbles and 2 white marbles in a bag. What
   is the probability of choosing a black marble, not replacing it,
   then choosing a white marble?

   ____

Holt Mathematics

Practice B

Probability of Independent and Dependent Events

Decide if each set of events is independent or dependent. Explain your answer.

1. A student spins a spinner and chooses a Scrabble® tile

____

____

2. A boy chooses a sock from a drawer of socks, then chooses a second sock without replacing the first.

____

____

3. A student picks a raffle ticket from a box, replaces the ticket, then picks a second raffle ticket.

____

____

Find the probability of each set of independent events.

4. drawing a red checker from a bag of 9 black checkers and 6 red checkers, replacing it, and drawing another red checker

____

5. drawing a black checker from a bag of 9 black checkers and 6 red checkers, replacing it, and drawing a red checker

____

6. rolling a 1, 2, or 3 on the first roll of a 1–6 number cube and rolling a 4, 5, or 6 on the second roll of the same cube

____

Solve.

7. Randy has 4 pennies, 2 nickels, and 3 dimes in his pocket. If he randomly chooses 2 coins, what is the probability that both are dimes?

____

Holt Mathematics

<table><tr><td>LESSON
11-5</td><td>

Practice C
Probability of Independent and Dependent Events
</td></tr></table>

Decide if each set of events is independent or dependent. Explain your answer.

1. A student spins an even number on a spinner and then spins another even number on the second spin.

   ____

   ____

2. A student guesses on two multiple choice questions.

   ____

   ____

Find the probability of each set of independent events.

3. drawing a brown sock from a drawer of 4 brown socks, 10 black socks, and 6 gray socks, replacing it, then drawing a gray sock

   ____

4. drawing a pair of black socks from a drawer of 4 brown socks, 10 black socks, and 6 gray socks

   ____

Solve.

5. Calista has 4 one-dollar bills, 2 five-dollar bills, and 3 ten-dollar bills in her wallet. If she randomly chooses 2 bills from her wallet, what is the probability that both are five dollar bills?

   ____

6. There are 10 true/false questions on a test. You do not know the answer to 4 of the questions, so you guess. What is the probability that you will get all 4 answers right?

   ____

7. There are 12 students in the After-School Club. Three names are being drawn at random to help with the school carnival. What is the probability that Jess, Sandy, and Phil will be chosen?

   ____

Holt Mathematics

Name _________________________________ Date __________ Class __________

Events are *independent* when the outcome of one event has no effect on the outcome of a second event. Rolling a number cube and flipping a coin are **independent events.**

Find the probability of rolling a 4 and flipping heads.

1. There are ______ outcomes for the number cube and ______ outcomes for the coin.

2. Using the Fundamental Counting Principle, there are

   ______ × ______, or ______, possible outcomes of rolling a number cube and flipping a coin.

3. Make a list of the possible outcomes:

   ____

   ____

   How many possible ways are there of rolling a 4 and flipping heads? ______

4. $P(4 \text{ and heads}) = \dfrac{\text{number of ways 4 and heads can occur}}{\text{number of possible outcomes}}$ = ______

Events are *dependent* when the outcome of one event does have an effect on the outcome of the next event. Drawing two marbles in a row from a bag without replacing the first marble are **dependent events.**

A bag contains 3 blue and 5 red marbles. Find the probability of drawing 2 blue marbles in a row without replacing the first marble.

5. The total number of marbles in the bag is ______. There are ______ blue

   marbles in the bag. $P(\text{blue marble on first draw})$ = ______

6. There are now ______ marbles in the bag. If you drew a blue marble on the first

   draw, there are ______ blue marbles left in the bag.

   $P(\text{blue marble on second draw})$ = ______

7. $P(\text{blue, blue}) = P(\text{blue on 1st draw}) \times P(\text{blue on 2nd draw}) =$ ___ × ___ = ___

Holt Mathematics

Challenge

LESSON 11-5

Pascal's Triangle

A special pattern, called Pascal's Triangle, can be used to find some probabilities. The triangle is called Pascal's Triangle because Pascal was one of the first mathematicians to formally study probability.

```
Row
 0                    1
 1                 1     1
 2              1     2     1
 3           1     3     3     1
 4        1     4     6     4     1
 5     1     5    10    10     5     1
```

Suppose 3 coins are flipped.

1. List all the possible outcomes.

____

2. How many outcomes show: all heads? 2 heads and 1 tail?
2 tails and 1 head? 3 tails?

____

3. How do the outcomes compare to row 3 of Pascal's Triangle?

____

Find the probability of each event when flipping 3 coins.

4. 3 heads? _________

5. 2 heads? _________

6. 1 head? _________

7. 0 heads? _________

Use Pascal's Triangle to find the following probabilities if 4 coins are flipped.

8. 4 tails _________

9. 3 tails _________

10. 2 tails _________

11. 1 tail _________

12. 0 tails _________

13. 2 heads _________

Holt Mathematics

LESSON 11-5 **Problem Solving**
Probability of Independent and Dependent Events

Write the correct answer.

1. Li rolls a pair of number cubes twice. On both rolls, the sum is 7. Are the rolls dependent or independent events?

2. Nine boys and 12 girls want to play soccer. Teams are formed by selecting one player at a time. Is the probability of selecting a boy after a girl is selected a dependent or an independent event?

3. Gregg has 12 cards. Half are black, and half are red. He picks two cards out of the deck. What is the probability that both cards are red?

4. In basketball, Alan makes 1 out of every 4 free throws he attempts. What is the probability that Alan will make his next 3 free throws?

Choose the letter for the best answer.

5. There are 8 blue marbles and 7 red marbles in a bag. Julie pulls two marbles at random from the bag first. What is the probability that she first pulls a blue marble and then a red marble?

A $\frac{8}{15}$ C $\frac{4}{7}$

B $\frac{4}{15}$ D $\frac{1}{2}$

6. You roll a 1–6 number cube twice. What is the probability that you roll a 3 on the first roll and a 6 on the second roll?

F $\frac{1}{36}$ H $\frac{1}{6}$

G $\frac{1}{9}$ J $\frac{1}{2}$

7. Andrew has $2.00 in quarters in his pocket, including three state quarters. He takes two quarters out of his pocket. What is the probability that they are **not** state quarters?

A $\frac{3}{8}$ C $\frac{3}{14}$

B $\frac{5}{8}$ D $\frac{5}{14}$

8. Jamie has 3 raffle tickets. One hundred tickets were sold. Her name was not drawn for the first prize. What is the probability that her name will be drawn for the second prize?

F $\frac{1}{3}$ H $\frac{1}{33}$

G $\frac{3}{100}$ J $\frac{2}{99}$

Holt Mathematics

Reading Strategies
LESSON 11-5 *Focus on Vocabulary*

When the outcome of one event does not affect the probability of the outcome of another, they are called **independent events**.

If you flip a coin once, it can land on head or tails.

$P(\text{landing on tails}) = \dfrac{1}{2}$

If you flip a coin a second time, $P(\text{landing on tails})$ is still $\dfrac{1}{2}$.

When the outcome of one event affects the probability of the outcome of another event, the events are called **dependent events**.

There are 52 cards in a deck, and 4 aces in the deck.

$P(\text{drawing an ace}) = \dfrac{4}{52}$

If an ace is drawn and not replaced, the probability of drawing an ace on the second draw drops to 3 out of 51.

$P(\text{drawing another ace}) = \dfrac{3}{51}$

Write *dependent events* or *independent events* to complete each sentence.

1. You roll a cube. Then you roll the cube a second time.

   ____

2. You take a coin from the jar and do not replace it. Then you take another coin from the jar.

   ____

3. One person in the class is chosen to be first in line. Another person is chosen to turn off the lights.

   ____

4. You have a bag with red, white, and blue counters. Explain how to conduct an experiment pulling counters from the bag so you will have independent events.

   ____

   ____

5. Explain how to change the experiment so the events are dependent.

   ____

   ____

Holt Mathematics

LESSON 11-5

Puzzles, Twisters & Teasers
Furry Math!

Decide whether each event is dependent or independent. Circle the letter above your answer. Match the letters with the number of the problems to solve the riddle.

1. tossing a coin twice

   Q   **T**

   dependent   independent

2. spinning a spinner five times

   M   **H**

   dependent   independent

3. pulling two socks from a drawer at the same time

   E   **P**

   dependent   independent

4. drawing two marbles out of a bag at the same time

   O   **D**

   dependent   independent

5. throwing a pair of dice ten times

   X   **U**

   dependent   independent

6. drawing two names out of a hat without replacement

   T   **B**

   dependent   independent

7. picking cards from a deck without replacement

   S   **V**

   dependent   independent

8. spinning two different spinners two times

   K   **I**

   dependent   independent

9. throwing three coins three times

   A   **D**

   dependent   independent

10. throwing one die five times

    U   **E**

    dependent   independent

Which side of a rabbit has the most fur?

___ ___ ___ ___ ___ ___ ___ ___ ___ ___
1   2   3   4   5   6   7   8   9   10

Practice A

LESSON 11-6

Combinations

Solve each problem in Column A. Draw a line to the correct answer in Column B.

Column A	Column B

1. If you have watermelon, cantaloupe and honeydew, how many combinations of two fruits are there?

A. 20

2. How many three-letter combinations are possible using the letters D, E, F, G, and H?

B. 15

3. How many different two-person debating teams can be chosen from six students?

C. 21

4. On Mondays at Pizza Pan, you can choose two toppings at no extra cost for your pizza. The toppings are mushrooms, sausage, peppers, and extra cheese. How many different pizzas could you order?

D. 10

5. How many different three-person relay teams are possible with four people?

E. 3

6. The students in Mr. Trumbull's English class need to read 2 books from a list of 7 books. How many different combinations of books are possible?

F. 6

7. Erin has 6 colors of ribbon: red, white, gold, green, blue, and purple. She makes bows of out 3 different colors of ribbon. How many different combinations of colors can she choose?

G. 4

Holt Mathematics

LESSON 11-6

Practice B
Combinations

1. A chef has some broccoli, cauliflower, carrots, and squash to make a vegetarian dish. List the possible combinations if he uses only 3 vegetables in the dish.

   ____

   ____

2. Lauren, Manuel, Nick, Opal, and Pat are forming groups of two to work on a drama production. List the different combinations of students that are possible using the first initial of each name.

   ____

3. Keiko has seven colors of lanyard. She uses three different colors to make a key chain. How many different combinations can she choose?

   ____

4. On Sundays at Ice Cream Heaven, you can choose two free toppings for your sundae. The toppings are nuts, hot fudge, caramel, and sprinkles. How many different combinations of toppings can you order?

   ____

5. How many different three-person relay teams can be chosen from six students?

   ____

6. The students in Mrs. Mandel's class need to choose two class representatives from six nominated students. How many different combinations of class representatives are possible?

   ____

7. There are four varieties of muffins available at the Coffee Shop. How many different ways can you choose three different muffins?

   ____

8. How many two-person carpools are possible with seven people?

   ____

Holt Mathematics

 ## Practice C
Combinations

1. If you have tomatoes, red peppers, onions, green peppers, and mushrooms, how many combinations of three vegetables are there? _______________

2. How many three-letter combinations are possible from P, Q, R, S, T, and U? _______________

3. Kim has seven colors of hair ribbons: red, white, pink, green, blue, black, and purple. She uses two different colors to make a bow. How many different combinations of colors can she choose? _______________

4. Les, Dennis, Greg, and Philip are pairing up to play doubles tennis. In how many different ways can they pair up? _______________

5. You can choose two side dishes for your chicken dinner. The choices are mashed potatoes, roasted potatoes, squash, carrots, string beans, stuffing, broccoli, or corn. How many different side dish combinations could you order? _______________

6. How many different five-person relay teams can be chosen from seven students? _______________

7. The students in Mr. Cohen's science class need to complete three experiments at home from a list of seven topics. How many different combinations of experiments are possible? _______________

8. Lena, Drew, Michaela, Trey, Nancy, and Jonah are pairing up to play one-on-one basketball. In how many different ways can they pair up? _______________

Find the number of combinations.

9. 8 things taken 4 at a time

____

10. 6 things taken 4 at a time

____

11. 8 things taken 5 at a time

____

12. 7 things taken 6 at a time

____

13. 7 things taken 4 at a time

____

14. 8 things taken 6 at a time

____

Holt Mathematics

LESSON 11-6 Reteach
Combinations

A **combination** is a selection of objects in which the order is not important. You can make a list to find the number of combinations.

The school district is going to plant two types of trees around the playground. The landscaper has five kinds of trees to choose from.

Let the letters A, B, C, D, and E represent the different kinds of trees. You can make an organized list of all possible combinations.

Tree **A** with each other tree: **AB** **AC** **AD** **AE**
Tree **B** with each other tree: **BA** **BC** **BD** **BE**
Tree **C** with each other tree: **CA** **CB** **CD** **CE**
Tree **D** with each other tree: **DA** **DB** **DC** **DE**
Tree **E** with each other tree: **EA** **EB** **EC** **ED**

1. How many groups of 2 trees are in the list? ______

2. AB is the same combination as ______.   BD is the same combination as ______.

   AC is the same combination as ______.   BE is the same combination as ______.

   AD is the same combination as ______.   CD is the same combination as ______.

   AE is the same combination as ______.   CE is the same combination as ______.

   BC is the same combination as ______.   DE is the same combination as ______.

3. Cross out each of the duplications in the list above. How many are left? ______

4. Make an organized list of all possible two-letter combinations using the letters U, V, W, X, Y, and Z. Cross out each duplication.

   UV, ______, ______, ______, ______   VU, ______, ______, ______, ______

   WU, ______, ______, ______, ______   XU, ______, ______, ______, ______

   YU, ______, ______, ______, ______   ZU, ______, ______, ______, ______

   There are ______ combinations of 6 letters taken 2 at a time.

Holt Mathematics

Name ____ Date ___________ Class ___________

Challenge
You Can Count on It!

You can use the Fundamental Counting Principle to help you find a formula that gives the number of combinations that are possible when *x* objects are taken *n* objects at a time. Remember, the Fundamental Counting Principle states that if a first event can occur in *a* ways, and a second event can occur in *b* ways, then the two events can occur together in *a · b* ways.

Suppose there are six students. How many different teams of three students can be chosen?

1. Use the Fundamental Counting Principle to find how many ways three of the students can be chosen for a team.

Choices for the first student	Choices left for the second student	Choices left for the third student	Total number of ways

_________ × _________ × _________ = _______ ways

2. Use the Fundamental Counting Principle to find how many ways any group of three students can be ordered.

Choices for the first student	Choices left for the second student	Choices left for the third student	Total number of orders

_________ × _________ × _________ = _______ orders

3. The order in which the three students are chosen for the team does not matter, because changing the order does not change the team. So, to find the total number of teams that are possible, divide the number of ways three students can be chosen (Problem 1) by the total number of orders (Problem 2).

_______ ways ÷ _______ orders = _______ teams

Use the Fundamental Counting Principle and the formula, Number of Combinations = Number of Ways ÷ Number of Orders, to find the number of combinations.

4. 9 things taken 5 at a time

____

5. 12 things taken 3 at a time

____

6. 10 things taken 4 at a time

____

7. 12 things taken 4 at a time

____

Holt Mathematics

LESSON 11-6 **Problem Solving**
Combinations

Write the correct answer.

1. Six friends are going to play a ball game. Each team has 3 players. How many different team combinations are possible?

2. Yung wants to visit 3 of the 5 Great Lakes this summer. How many different combinations of lakes are possible?

3. There are 4 spots left for a school field trip. Roberta, Samuel, Thea, Ling, Jose, and Mark all want to go on the trip. How many different combinations are possible for the remaining 4 spots?

4. At summer camp, the campers pick two activities for the first day of camp. They can choose from among hiking, canoeing, rock climbing, and bird watching. How many different combinations of activities are there?

Choose the letter for the best answer.

5. Eight children are playing a trivia game. They want to make teams of 2 players each. How many different team combinations are possible?

A 10 combinations
B 21 combinations
C 28 combinations
D 56 combinations

6. At Washington Middle School, a student takes 4 core classes a day. There are 6 different core classes offered. How many different combinations of classes are there?

F 10 combinations
G 12 combinations
H 15 combinations
J 24 combinations

7. The class is drawing maps of the 7 continents to display in the school lobby. The main wall in the lobby has room for 3 maps. How many combinations of maps are possible for that location?

A 10 combinations
B 21 combinations
C 30 combinations
D 35 combinations

8. Ms. Henrie's literature class is voting on the 2 most influential people of the eighteenth century. The choices are 5 famous people. How many different combinations are possible?

F 7 combinations
G 8 combinations
H 10 combinations
J 15 combinations

48

Holt Mathematics

LESSON 11-6 Reading Strategies
Make an Organized List

Here are three different frozen yogurt flavors: vanilla, chocolate, strawberry. How many **combinations** of only two different flavors can you make?

Order does not matter. If you choose a scoop of vanilla first and a scoop of chocolate second, it is the same combination as choosing chocolate first and then vanilla.

Making a list can help you picture the possible combinations of two different flavors.

Scoop 1	Scoop 2
vanilla	chocolate
vanilla	strawberry
?	?

Use the chart to answer each question.

1. What flavor can you pair with strawberry to make a new combination?

   ____

2. Does pairing a scoop of strawberry with a scoop of vanilla make a new combination? Why or why not?

   ____

3. How many different combinations of two different flavors of yogurt can you make with three flavors?

   ____

4. Assume banana is added as another choice. Fill in the table to list all the combinations that can be made using two different flavors.

   ____

Scoop 1	Scoop 2	Scoop 1	Scoop 2

Holt Mathematics

Puzzles, Twisters & Teasers

LESSON 11-6 *Banana-Rama!*

Find the number of possible combination of pairs for each set of items below. Match the letters with the correct answers to solve the riddle.

1. peas, potatoes, corn, carrots, beans

   combinations: ______ **A**

2. Call of the Wild, Treasure Island, Peter Pan

   combinations: ______ **E**

3. bananas, apples

   combinations: ______ **N**

4. N'Sync, Britney Spears, Backstreet Boys, Janet Jackson

   combinations: ______ **X**

5. chocolate, vanilla, strawberry, pineapple, butter pecan, toffee, mint, blueberry, raspberry, cherry

   combinations: ______ **B**

6. Rob, Karen, Sari, George, Lemuel, Jenny, Tom, Lesley

   combinations: ______ **C**

7. blue shirt, black shirt, yellow shirt, white shirt, green shirt, brown shirt, gray shirt

   combinations: ______ **H**

8. Winston Churchill

   combinations: ______ **P**

9. Monet, Manet, Renoir, Pissarro, Degas, Cassatt, Picasso, Van Gogh, Gauguin

   combinations: ______ **T**

10. tennis ball, golf ball, football, soccer ball, basketball, snow ball

    combinations: ______ **S**

How is a flamingo like a bunch of bananas?

They're both pink, ___ ___ ___ E ___ T
                     3   6   28   0

___ ___ E ___ A ___ ___ N A ___
36 21   45   1   10   15 .

Holt Mathematics

Practice A
Permutations

1. In how many ways can you arrange the letters in the word NOW? List the permutations.

____

2. In how many ways can you arrange the numbers 4, 5, 6, and 7 to make a three-digit number? List the permutations.

____

____

3. Find the number of permutations of the letters in the word FOUR. _________________

4. In how many ways can you arrange the numbers 3, 4, 5, 6, and 7 to make a five-digit number? _________________

5. Find the number of ways you can arrange the letters in the word *arrange*. _________________

Choose the letter for the best answer.

6. What is another way of showing 5 factorial or 5!?

   A 5^3

   B $5 + 4 + 3 + 2 + 1$

   C $5 \cdot 4 \cdot 3 \cdot 2 \cdot 1$

   D 5^5

7. How many permutations of the numbers 10 through 14 are there?

   F 10!

   G 5!

   H 4!

   J 14!

8. In how many ways can 5 children be matched with 5 puppies?

   A 5   C 100

   B 25   D 120

9. Six friends are waiting in line at the movie theater. In how many different orders can they be standing in line?

   F 720   H 36

   G 120   J 6

10. How many permutations of the numbers 1 through 9 are there?

    A 9   C 9!

    B 900,000   D 9^9

11. In how many different ways can seven drivers be matched up with seven rental cars?

    F 5,040   H 49

    G 2,401   J 7

Holt Mathematics

<table><tr><td>LESSON
11-7</td><td>

Practice B
Permutations
</td></tr></table>

1. Joe has homework assignments for math, Spanish, and history. In how many different orders can he do his homework?   _______________

2. Find the number of permutations of the letters in the word SMART.   _______________

3. In how many ways can you arrange the numbers 6, 7, 8, and 9 to make a four-digit number?   _______________

4. A table has 8 seats. In how many different ways can 8 people sit at the table?   _______________

5. Nine mountain bikers are on a bicycle trip. In how many possible ways can they follow each other?   _______________

6. Seven students are waiting in line at the cafeteria. In how many different orders can they be standing in line?   _______________

7. How many permutations of the letters A through F are there?   _______________

8. Ed, Martine, Sal, Carl, Paula, Terry, Ken, Leo, Ursula, and Jamie are in a race. In how many different orders can they finish?   _______________

9. Find the number of permutations of the letters in the word *permutations.*   _______________

10. In how many different orders can 11 people stand in line?   _______________

11. In how many different ways can a librarian arrange eight books on a shelf?   _______________

12. Melinda has 15 art trophies. Write an expression that shows how many different ways she can line up her trophies on a shelf.   _______________

Holt Mathematics

Practice C
Permutations

1. **a.** Find the number of permutations of the letters
   in the word BRAIN. _______________

   b. If you choose one of the permutations at random,
   what is the probability that it will start with a vowel? _______________

2. **a.** In how many ways can you arrange the numbers
   4, 5, 6, 7, 8, and 9 to make a six-digit number? _______________

   b. If you choose one of the six-digit numbers at
   random, what is the probability that the number
   will be less than 600,000? _______________

3. Eight skydivers are on an airplane. In how many
   possible ways can they jump from the plane? _______________

4. There are seven different flavors of yogurt at a
   supermarket. In how many different orders can
   they be lined up on the shelf? _______________

5. How many permutations of the numbers 1 through
   9 are there? _______________

6. In how many different ways can a video store
   arrange ten videos on a shelf? _______________

7. Harrison has a collection of 25 antique teapots.
   Write an expression that shows in how many
   different ways he can display his teapots. _______________

**Determine whether each problem involves permutations
or combinations.**

8. Elect four people to be president, vice president,
   secretary, and treasurer. _______________

9. Order a three-topping pizza from a choice of
   eight toppings. _______________

10. Decide how many ways you can put ten books
    on a shelf. _______________

11. Organize 30 students into two-person doubles teams. _______________

Holt Mathematics

Reteach
LESSON 11-7
Permutations

A **permutation** is a selection of objects in a particular order.

In how many ways can Allie, Bob, and Carl stand in a line?

You can draw a tree diagram to find the number of permutations.

1. Complete the tree diagram.

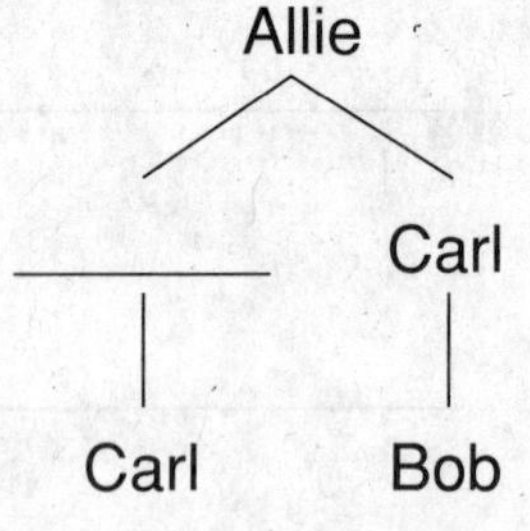
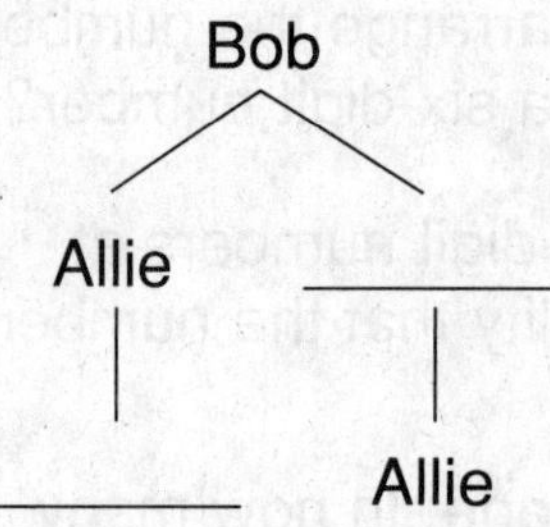
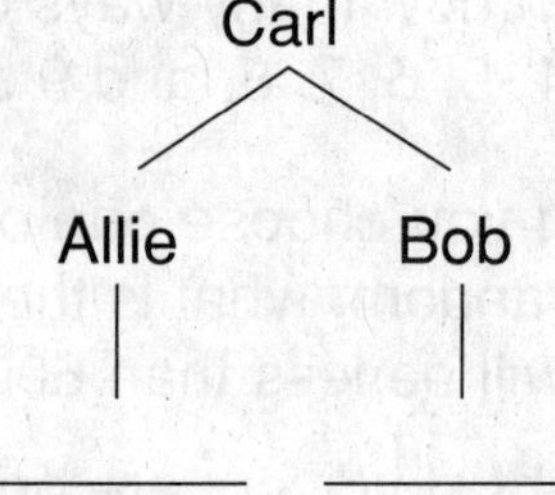

2. Complete the list of the outcomes.

__________ Bob Carl   Bob __________ Carl   Carl Allie __________

Allie Carl __________   Bob __________ Allie   __________ Bob Allie

You can also use multiplication to find the number of permutations.

3. Complete the multiplication.

_____ × _____ × _____ = _____

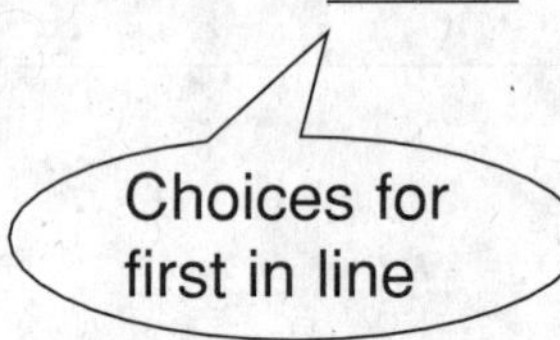

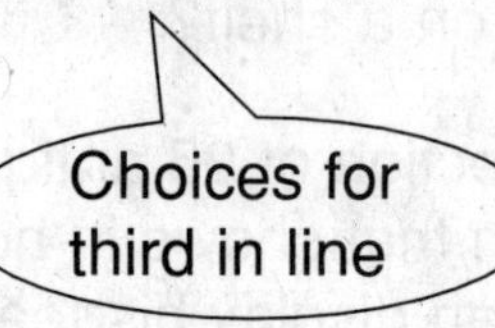

4. There are _____ permutations for 3 people standing in line.

You can find the probability that they will be standing in line alphabetically.

5. P(alphabetical order) = $\dfrac{\text{number of alphabetical arrangements}}{\text{total number of permutations}}$ = __________

6. Find the number of ways you can arrange the letters in the word MATH.

_____ • _____ • _____ • _____ = __________

Holt Mathematics

Challenge
Factorial Fun

Sometimes you can use the definition of *factorial* to simplify computations with factorials.

Example: $\dfrac{8!}{5!2!} = \dfrac{8 \cdot 7 \cdot {}^{3}6 \cdot 5 \cdot 4 \cdot 3 \cdot 2 \cdot 1}{(5 \cdot 4 \cdot 3 \cdot 2 \cdot 1)(2 \cdot 1)} = \dfrac{8 \cdot 7 \cdot 3}{1} = 168$

Simplify. Remember to use the order of operations.

1. $(3 + 2)!$ **2.** $(4!)(2!)$ **3.** $(2 \cdot 3)!$ **4.** $5 + 4!$

____

5. $\dfrac{6!}{7!}$ **6.** $\dfrac{6!}{4!}$ **7.** $\dfrac{5!}{7!}$ **8.** $\dfrac{9!}{7!}$

____

9. $\dfrac{8!}{3!5!}$ **10.** $\dfrac{6!}{2!4!}$ **11.** $\dfrac{9!}{8!3!}$ **12.** $\dfrac{10!}{8!4!}$

____

13. $\dfrac{8!}{4!4!}$ **14.** $\dfrac{3!5!}{4!}$ **15.** $\dfrac{11!}{7!4!}$ **16.** $\dfrac{6!8!}{7!5!}$

____

17. $\dfrac{2! + 2!}{3!}$ **18.** $\dfrac{3!3! + 3}{3!}$ **19.** $\dfrac{4!}{3! + 4}$ **20.** $3\left(\dfrac{3!}{4!}\right)$

____

21. $2!(3!)$ **22.** $4! - 3!$ **23.** $4!(3!)(2!)$ **24.** $5 + 4! + 3$

____

Holt Mathematics

Problem Solving

LESSON 11-7 *Permutations*

Write the correct answer.

1. Five snowboarders are competing in a half-pipe competition during the Winter Sports Festival. In how many different orders can the snowboarders compete?

2. In how many different orders can, Debbie, Brigitte, and Adam wait in line in the school cafeteria? What is the probability that they will be in alphabetical order?

3. In how many different orders can the science class study the planets Jupiter, Saturn, Uranus, and Neptune? What is the probability that they will study Saturn first?

4. The physical education teacher sets up 8 different exercise stations for the class to complete. In how many different orders can the stations be done?

Choose the letter for the best answer.

5. Hannah, Javier, and Beth were the three qualifiers for a race. In how many different orders can they finish? What is the probability that Hannah or Beth will be first?

A 3 orders; $\frac{1}{3}$

B 3 orders; $\frac{2}{3}$

C 6 orders; $\frac{1}{3}$

D 6 orders; $\frac{2}{3}$

6. Berto and 5 friends have ordered dinner at a restaurant. What is the probability that Berto will be served last?

F $\frac{1}{720}$

G $\frac{1}{120}$

H $\frac{1}{60}$

J $\frac{1}{6}$

7. Six different prizes are being awarded to the winners of a contest. In how many different ways can the prizes be awarded?

A 4,320 ways   **C** 120 ways

B 720 ways   **D** 30 ways

8. Jonathan will have math, English, history, social studies, and science each day next year. What is the probability that his classes will be in alphabetical order?

F $\frac{1}{5}$ **H** $\frac{1}{120}$

G $\frac{1}{24}$ **J** $\frac{1}{720}$

Holt Mathematics

Reading Strategies
LESSON 11-7 *Make an Organized List*

An arrangement of objects in a certain order is called a
permutation.

How many different ways are there to arrange turkey, lettuce, and
tomato on a sandwich?

You can make a list of the items just as they would appear on the
sandwich. If turkey is on top, the lettuce and tomato could be
arranged as follows:

Turkey   Turkey
↓   ↓
Lettuce   Tomato
↓   ↓
Tomato   Lettuce

Answer each question.

1. List the ways you can arrange the three items with lettuce on
   the top.

   ____

2. How many different arrangements are there with lettuce on top?

   ____

3. List the ways you can arrange the three items with the tomato
   on top.

   ____

4. Make an organized list showing all the different arrangements
   that can be made with turkey, lettuce, and tomato.

   ____

   ____

5. How many different ways can you arrange tomato, lettuce and
   turkey on a sandwich?

   ____

Holt Mathematics

LESSON 11-7

Puzzles, Twisters & Teasers
Stay on Track!

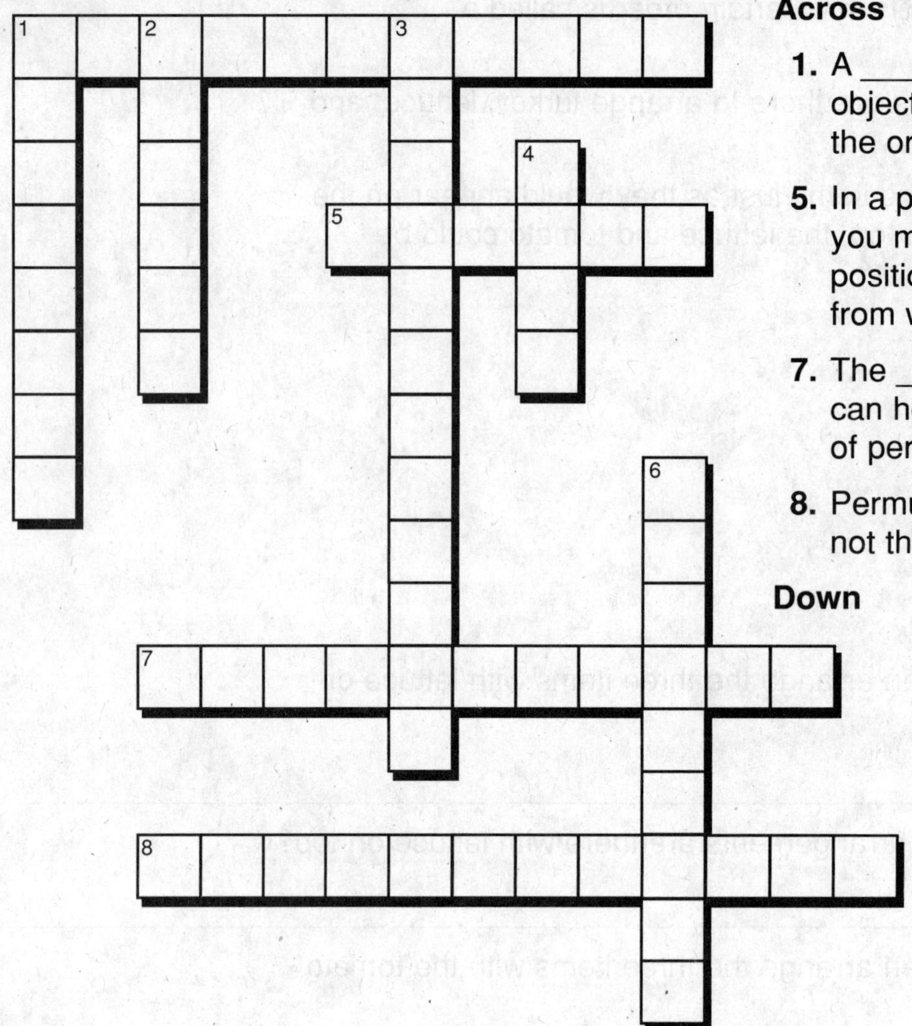

Across

1. A _____ is an arrangement of objects or events in which the order is important.

5. In a permutation, each time you make a _____ for one position, there is one less from which to choose.

7. The _____ Counting Principle can help you find the number of permutations.

8. Permutations and _____ are not the same.

Down

1. The number of _____ permutations of a group of objects may be written as a factorial.

2. The order of a group of objects may be chosen at _____.

3. An _____ permutation is based on the letters in the names of the objects.

4. You can use a _____ to find the number of permutations of a group of objects.

6. You can find the _____ of 6 by multiplying 6 × 5 × 4 × 3 × 2 × 1.

58

Holt Mathematics

LESSON 11-1 — Practice A — Probability

Match each event to its likelihood.

1. rolling a number greater than 6 on a number cube labeled 1 through 6 __D__   A likely

2. flipping a coin and getting heads __C__   B unlikely

3. drawing a red or blue marble from a bag of red marbles and blue marbles __E__   C as likely as not

4. spinning a number less than 3 on a spinner with 8 equal sections marked 1 through 8 __B__   D impossible

5. rolling a number less than 6 on a number cube labeled 1 through 6 __A__   E certain

Solve.

6. A bag contains 4 red marbles, 3 green marbles, and 2 yellow marbles. The probability of randomly picking a yellow marble is $\frac{2}{9}$. What is the probability of not picking a yellow marble? __$\frac{7}{9}$__

7. A number cube is labeled 1 through 6. The probability of randomly rolling a 4 is $\frac{1}{6}$. What is the probability of not rolling a 4? __$\frac{5}{6}$__

Tell whether the event is impossible, unlikely, as likely as not, likely, or certain.

8. Janelle almost never eats meat. On Monday, the school cafeteria offers three main choices. The choices are hamburger, tuna, or a turkey sandwich. Estimate the probability that Janelle will choose a hamburger. __unlikely__

9. Tyrone rides his bicycle to school if he gets up by 7:15 A.M. Tyrone gets up by 7:15 A.M. about half the time. Estimate the probability that Tyrone will ride his bicycle to school. __as likely as not__

3   **Holt Mathematics**

LESSON 11-1 — Practice B — Probability

Determine whether each event is impossible, unlikely, as likely as not, likely, or certain.

1. rolling an even number on a number cube labeled 1 through 6 __as likely as not__

2. picking a card with a vowel on it from a box of cards in which each letter of the alphabet is written on a card __unlikely__

3. spinning a number greater than 2 on a spinner with 10 equal sections marked 1 through 10 __likely__

4. drawing a red marble from a bag of black, blue, and green marbles __impossible__

5. flipping a coin and getting heads or tails __certain__

6. rolling a number that is less than three 5 times in a row on number on a number cube labeled 1 through 6 __unlikely__

Solve.

7. A bag contains 3 green marbles, 7 blue marbles, and 2 black marbles. The probability of randomly picking a green marble is $\frac{1}{4}$. What is the probability of not picking a green marble? __$\frac{3}{4}$__

8. A spinner has 8 equal sections labeled 1 through 8. The probability of spinning a number that is greater than or equal to 6 is $\frac{3}{8}$. What is the probability of spinning a number that is not greater than or equal to 6? __$\frac{5}{8}$__

9. The probability of randomly drawing a red card from a bag that contains red, blue, and green cards is $\frac{3}{10}$. What is the probability of not drawing a red card? __$\frac{7}{10}$__

10. Myra almost always spends at least 45 minutes on the treadmill. If Myra got on the treadmill at 5:20 P.M., estimate the probability that she will still be on the treadmill at 6:00. __likely__

11. Morris rarely arrives home before 4:00 P.M. It is now 3:20 P.M. Estimate the probability that Morris will arrive home in the next 30 minutes. __unlikely__

4   **Holt Mathematics**

LESSON 11-1 — Practice C — Probability

Determine whether each event is impossible, unlikely, as likely as not, likely, or certain.

1. rolling a number less than 4 on a number cube labeled 1 through 6 __as likely as not__

2. picking a card with a multiple of 3 from a box with 10 number cards numbered 1 through 10 __unlikely__

3. drawing a red marble from a bag of 3 blue marbles, 8 red marbles, and 2 green marbles __likely__

4. rolling a number cube that has sides labeled 2, 4, 6, 8, 10, and 12, and getting an even number __certain__

5. picking two cards that have a sum of greater than 10 from a set of number cards in which there are three 4's, five 3's, and two 5's __impossible__

6. drawing a yellow marble from a bag of 5 blue marbles, 2 green marbles, and 7 yellow marbles __as likely as not__

Solve.

7. A spinner has 16 sections labeled 1 through 16. The probability of spinning a number that is less than or equal to 10 is $\frac{5}{8}$. What is the probability of spinning a number that is not less than or equal to 10? __$\frac{3}{8}$__

8. The probability of randomly picking a striped marble from a bag that contains red, blue, and striped marbles marbles is $\frac{5}{12}$. What is the probability of not picking a striped marble? __$\frac{7}{12}$__

9. When Sam studies for 1 hour or more, he almost always gets at least a B on his math test. Sam studies from 8:45 to 10:00. Estimate the probability that Sam will get a B on his quiz. __likely__

10. A bag contains 12 black cards and 11 red cards. Julia randomly draws 2 black cards and does not replace them. Will Julia be more likely to draw a black card than a red card on her next draw? Explain.

__No. Since the bag will now contain 10 black cards and 11 red cards, Julia will be more likely to pick a red card.__

5   **Holt Mathematics**

LESSON 11-1 — Reteach — Probability

You can describe the probability of an event as **impossible, unlikely, as likely as not, likely,** or **certain.**

For the spinner at the right:

- Spinning a 7 is **impossible** because the spinner has no 7.
- Spinning a 5 is **unlikely** because only 1 of the 6 numbers is a 5.
- Spinning an even number is **as likely as not** because 3 of the numbers are even and 3 of the numbers are odd.
- Spinning a number that is greater than 1 is **likely** because 5 of the 6 numbers are greater than 1.
- Spinning a number that is less than 7 is **certain** because all of the numbers are less than 7.

Use the spinner. Write impossible, unlikely, as likely as not, likely, or certain to complete each statement.

1. Four of the 6 numbers are less than 5.
Spinning a number that is less than 5 is ___likely___.

2. Six of the 6 numbers are greater than 0.
Spinning a number that is greater than 0 is ___certain___.

3. One of the 6 numbers is a 3.
Spinning a 3 is ___unlikely___.

A bag contains 1 red marble, 2 blue marbles, and 3 green marbles.

The probability of picking a red marble is $\frac{1}{6}$.
To find the probability of not picking a red marble,
subtract the probability of picking a red marble from 1. $P = 1 - \frac{1}{6} = \frac{5}{6}$

Solve.

4. A number cube is labeled 1 through 6. The probability of randomly rolling a 3 is $\frac{1}{6}$. What is the probability of not rolling a 3? $P = 1 - \frac{1}{6} = \frac{5}{6}$

5. A spinner has 5 sections labeled 1 through 5. The probability of randomly spinning an even number is $\frac{2}{5}$. What is the probability of not spinning an even number? $P = 1 - \frac{2}{5} = \frac{3}{5}$

6   **Holt Mathematics**

Challenge
A Likely Choice

There are 20 cards in a deck. One card is randomly selected.
Provide additional information about the cards to describe each
type of outcome.

Example: Describe a situation with a likely outcome.
There are 2 face cards and 18 number cards in the deck.
It is likely that you will select a number card. **Answers may vary. Possible answers are given.**

1. Describe a situation with an unlikely outcome.

   There are 3 red cards and 17 black cards.

   It is unlikely that you will select a red card.

2. Describe a situation with a certain outcome.

   There are 20 red cards. It is certain that you

   will select a red card.

3. Describe a situation with an outcome that is as likely as not.

   There are 10 red cards and 10 black cards. It is

   as likely as not that you will select a black card.

4. Describe a situation with an impossible outcome.

   There are 20 number cards. It is impossible that

   you will select a face card.

5. A 1–6 number cube is rolled. Describe a roll with a likely
   outcome.

   Roll a number less than 6.

6. A 1–6 number cube is rolled. Describe a roll with an outcome
   that is as likely as not.

   Roll an even number.

7. A 1–6 number cube is rolled. Describe a roll with an impossible
   outcome.

   Roll a 10.

7

Holt Mathematics

Problem Solving
Probability

Write the correct answer.

1. Of the original 56 signers of the
   Declaration of Independence, four of
   them represented North Carolina. If
   you selected one signer randomly,
   how likely is it that he represented
   North Carolina?

   unlikely

2. One question on a social studies
   multiple-choice test has four possible
   answers. Marianne is sure two of the
   choices are incorrect. How likely is
   she to choose the correct answer?

   as likely as not

3. There are 8 right-handed pitchers
   and 2 left-handed pitchers on the
   Tigers baseball team. How likely is
   it that their opponents will face a
   right-handed pitcher?

   likely

4. Every seventh-grade student is
   attending a presentation about
   recycling in the auditorium. Jaleel is
   a seventh-grade student. How likely
   is it that he is in the auditorium?

   certain

Choose the letter for the best answer.

The table shows the contents of Leticia's
CD and DVD collection.

Leticia's CD and DVD Collection

Type of CD/DVD	Number
Rock CD	26
Pop CD	17
Comedy DVD	11
Drama DVD	2

5. How likely is it that a disk chosen
   randomly is a CD?
   - A unlikely
   - C certain
   - B as likely as not
   - (D) likely

6. How likely is it that a disk chosen
   randomly is a jazz CD?
   - F unlikely
   - H likely
   - G as likely as not
   - (J) impossible

7. Sally picks a disk at random from
   Leticia's collection. How likely is it
   that it is a comedy DVD?
   - A as likely as not
   - B likely
   - C certain
   - (D) unlikely

8. Leticia picks three disks from her
   collection at random. Which outcome
   is impossible?
   - F They are all rock CDs.
   - (G) They are all drama DVDs.
   - H None of them are rock CDs.
   - J They are all DVDs.

8

Holt Mathematics

Reading Strategies
Use Context

Probability measures how likely it is that an event will happen. Is it
likely that a coin will land on heads when it is tossed?

An outcome is the result of an experiment. Landing on heads is a
possible outcome of the experiment of tossing a coin.

Each outcome can be described in one of these ways:

• impossible • unlikely • as likely as not • likely • certain

Refer to the illustration for Exercises 1–5.

**Complete each sentence. Write *impossible, unlikely, as likely
as not, likely* or *certain*.**

1. It is   impossible   to draw a striped pebble from the bag.

2. You are   likely   to draw a pebble that is not black from
   the bag, because 9 of the 12 pebbles are not black.

3. If you reach into the bag, it is   certain   that you will draw
   a pebble.

4. Drawing a white pebble from the bag is   as likely as not   ,
   because 6 of the 12 pebbles are white.

5. Drawing a spotted pebble from the bag is   unlikely   ,
   because only 4 of the 12 pebbles are spotted.

Answer the following questions.

6. What word is used to measure the likelihood that an event will
   happen?   probability

7. What word means the result of an event?   outcome

9

Holt Mathematics

Puzzles, Twisters & Teasers
Put Up Your Ducks!

**Circle words from the list in the word search (horizontally,
vertically or diagonally). Find a word that answers the riddle
and write it on the line.**

trial   measure   probability   outcome   event
impossible   likely   unlikely   certain   experiment

```
I M P O S S I B L E P B E
N Q W E R T R I A L R O X
F A S D F G H J K L O U P
O Z X C V B N M I O B T E
R U N L I K E L Y U A B R
M E A S U R E V B N B O I
A P L M N K O I J B M M M
L I K E L Y F G H I L V E
B N C E R T A I N L I V N
U J I K O L E V E N T X T
A S D C H I C K E N Y G E
```

Why don't hens fight each other?

They're all _____ chicken _____

10

Holt Mathematics

Holt Mathematics

Practice A
Experimental Probability

Find the experimental probability in the box. Each answer can be used only once.

$\frac{4}{11}$	$\frac{7}{9}$	$\frac{11}{15}$	$\frac{2}{9}$	$\frac{4}{15}$	$\frac{7}{11}$

1. Jolene is playing basketball. She scores on 11 out of the 15 baskets she shoots.

   a. What is the experimental probability that Jolene will get a basket on the next shot? $\frac{11}{15}$

   b. What is the experimental probability that Jolene will not get a basket on the next shot? $\frac{4}{15}$

2. Jamie is playing baseball. He gets a hit 7 out of 9 times at bat.

   a. What is the experimental probability that Jamie will get a hit his next time at bat? $\frac{7}{9}$

   b. What is the experimental probability that Jamie will not get a hit his next time at bat? $\frac{2}{9}$

3. Lou Ann is practicing for an archery tournament. She hits the target 7 out of 11 times.

   a. What is the experimental probability that Lou Ann will hit the target on the next shot? $\frac{7}{11}$

   b. What is the experimental probability that Lou Ann will not hit the target on the next shot? $\frac{4}{11}$

Find the experimental probability. Write your answer as a fraction, as a decimal, and as a percent.

4. A batter gets 6 hits in 12 times at bat. What is the experimental probability that she will get a hit in her next time at bat? $\frac{1}{2}$; 0.5; 50%

5. A goalie blocks 16 out of 20 shots. What is the experimental probability that he will block the next shot? $\frac{4}{5}$; 0.8; 80%

Holt Mathematics

Practice B
Experimental Probability

Find the experimental probability. Write your answer as a fraction, as a decimal, and as percent.

1. Jaclyn is a soccer goalie. If she has 21 out of 25 saves in practice, what is the experimental probability that she will have a save on the next shot on goal? $\frac{21}{25}$; 0.84; 84%

2. If Harris hit the bull's-eye 3 out of 8 times at archery practice, what is the experimental probability that he will hit the bull's-eye on his next try? $\frac{3}{8}$; 0.375; 37.5%

3. Nathan inspects new pants at a factory. Of the first 56 pairs of pants he inspected 49 were acceptable. What is the experimental probability that the next pair of pants will be acceptable? $\frac{7}{8}$; 0.875; 87.5%

4. Sara has gone to work for 60 days. On 39 of those days she arrived at work before 8:30 A.M. On the rest of the days she arrived after 8:30 A.M. What is the experimental probability that she will arrive at work after 8:30 A.M. the next day she goes to work? $\frac{7}{20}$; 0.35; 35%

Solve:

5. After a movie premiere, 99 of the first 130 people surveyed said they liked the movie.

   a. What is the experimental probability that the next person surveyed will say he or she liked the movie? $\frac{99}{130}$

   b. What is the experimental probability that the next person surveyed will say he or she did not like the movie? $\frac{31}{130}$

6. For the past 30 days, Naomi has been recording the number of customers at her restaurant between 10 A.M. and 11 A.M. During that hour, there have been fewer than 20 customers on 25 out of 30 days.

   a. What is the experimental probability that there will be fewer than 20 customers on the thirty-first day? $\frac{5}{6}$

   b. What is the experimental probability that there will be more than 20 customers on the thirty-first day? $\frac{1}{6}$

7. For the past four weeks, Nestor has been recording the daily high temperatures. During that time, the high temperature has been below 45° on 20 out of 28 days. What is the experimental probability that the high temperature will be below 45° on the twenty-ninth day? $\frac{5}{7}$

Holt Mathematics

Practice C
Experimental Probability

Find the experimental probability. Write your answer as a fraction, as a decimal, and as a percent.

1. Luke is practicing his tennis serve. If he gets 21 out of 27 serves in, what is the experimental probability that he will get the next serve in? $\frac{7}{9}$; $0.\overline{7}$; 77.7%

2. Jose saw 50 people. Fourteen of them were wearing red shirts and 17 were wearing blue shirts. What is the experimental probability that the next person he sees will be wearing a blue shirt? $\frac{17}{50}$; 0.34; 34%

Solve.

3. During an exit survey after a play, 75 of the first 120 people surveyed said they did not like the play.

   a. What is the experimental probability that the next person surveyed will say he or she liked the play? $\frac{3}{8}$

   b. What is the experimental probability that the next person surveyed will say he or she did not like the play? $\frac{5}{8}$

4. For the past two weeks, Jimmy has been counting the number of joggers in the park between 8 P.M. and 9 P.M. each evening. In that time, there have been 40 or more joggers on 7 out of 14 days.

   a. What is the experimental probability that that there will be 40 or more joggers on the fifteenth day? $\frac{1}{2}$

   b. What is the experimental probability that that there will be fewer than 40 joggers on the fifteenth day? $\frac{1}{2}$

5. If Kathy hit the dartboard 9 out of 15 times and Toby hit the dartboard 14 out of 20 times, who has the greater experimental probability of hitting the dartboard on his or her next try? Toby

6. Mona works at a snack bar. Of the first 25 hot dogs ordered one day, 19 were ordered with sauerkraut. If the snack bar expects to sell 150 hot dogs on any day, how many would they expect to be ordered with sauerkraut? 114 hot dogs

Holt Mathematics

Reteach
Experimental Probability

Experimental probability is an estimate of the probability of an event. It is called *experimental* because you make observations or experiments to find the number of times a certain event actually happened.

Experimental probability $\approx \dfrac{\text{number of times a certain event happened}}{\text{total number of trials}}$

Suppose you have 12 marbles. Without replacing any marbles, you pull a red marble 5 times. What is the experimental probability of getting another red marble the next time you pull a marble?

$P(\text{red}) \approx \dfrac{\text{number of red marbles}}{\text{number of marbles}} = \dfrac{5}{12}$

The probability of getting red on the next pull is $\frac{5}{12}$.

At softball practice, Manny had 6 hits out of 15 times at bat.

1. What is the experimental probability that Manny will get a hit at his next time at bat?

   $P(\text{hit}) \approx \dfrac{\text{number of hits}}{\text{total number of times at bat}} = \dfrac{6}{15} = \dfrac{2}{5}$

2. What is the experimental probability that Manny will not get a hit at his next time at bat?

   $P(\text{no hit}) \approx 1 - \dfrac{\text{number of hits}}{\text{total number of times at bat}} = 1 - \dfrac{6}{15} = \dfrac{3}{5}$

Pam is playing darts. She hit the bull's eye 7 times out of 20 throws.

3. What is the experimental probability that Pam will hit the bull's eye on her next throw? $\frac{7}{20}$

4. What is the experimental probability that Pam will NOT hit the bull's eye on her next throw? $\frac{13}{20}$

So far this year Trisha's softball team has played 4 of their 20 games on Field A.

5. What is the experimental probability that they will play their next game on Field A? $\frac{1}{5}$

6. What is the experimental probability that they will play their next game on a field other than Field A? $\frac{4}{5}$

Holt Mathematics

Challenge — 11-2
Pondering Probability

Materials needed: paper plate, construction paper, metal fastener

Use a paper plate, a construction paper arrow, and a metal fastener to construct a spinner. Your spinner should have 6 or 8 congruent sections. Draw a different design in each section.

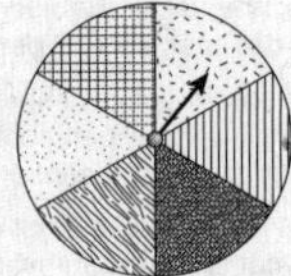

1. Based on the number of sections in your spinner, what fraction of the time would you predict that the pointer will stop on each section?

   Possible answers: $\frac{1}{6}$ for 6 sections;

   $\frac{1}{8}$ for 8 sections

Spin the pointer 50 times and record your results in the table below.

Section							
Times Spun							

2. What is your experimental probability of landing on each section?

   Answers will vary depending on results recorded

   in the table.

3. Was your experimental probability of landing on any section close to your prediction? Explain.

   Answers will vary; the experimental probability

   should be close to the prediction.

4. Were any of your results surprising? Explain.

   Answers will vary.

15
Holt Mathematics

Problem Solving — 11-2
Experimental Probability

Write the correct answer as a fraction in simplest form.

This table shows a breakdown by format of total music sales in the United States in 2004.

Total American Music Sales in 2004

Format	Total (% of units shipped)
CD	80
Digital Single	15
Music Video	3
Other	2

1. What is the experimental probability that any random music purchase in 2004 was a CD?

   $\frac{4}{5}$

2. What is the experimental probability that any random music purchase in 2004 was not a Music Video?

   $\frac{97}{100}$

3. What is the experimental probability that any random music purchase in 2004 was a digital single?

   $\frac{3}{20}$

4. Which combination of sales has an experimental probability of $\frac{1}{20}$?

   music videos and others

Choose the letter for the best answer.

5. Ethan hits 4 ringers in 10 attempts while pitching horseshoes. What does an experimental probability of $\frac{2}{5}$ describe?

   A P(horseshoes)
   B P(missed shots)
   C P(attempts)
   (D) P(ringers)

6. Jay beats Terry at table tennis 3 out of 5 games. What is the experimental probability that Terry will win their next game?

   F $\frac{1}{2}$
   (H) $\frac{2}{5}$
   G $\frac{3}{5}$
   J 1

7. Poonam counts 10 classmates out of 36 people in the library. What is the experimental probability that the next person will be a classmate?

   A $\frac{5}{36}$
   (B) $\frac{5}{18}$
   C $\frac{1}{36}$
   D $\frac{1}{10}$

8. Macy makes 15 of 20 free throws at basketball practice. What is the experimental probability that she will miss her next free throw?

   (F) $\frac{1}{4}$
   H $\frac{2}{3}$
   G $\frac{1}{2}$
   J $\frac{3}{4}$

16
Holt Mathematics

Reading Strategies — 11-2
Make Predictions

Experimental probability is a ratio. The ratio compares the number of times an event occurs to the total number of trials.

A trial is the number of times an experiment is carried out or an observation is made.

Experimental Probability

$$\text{probability} \approx \frac{\text{number of times an event occurs}}{\text{total number of trials}}$$

Refer to the illustration for Exercises 1 and 2. On this cube, you can land on 1, 2, or 3. Answer each question.

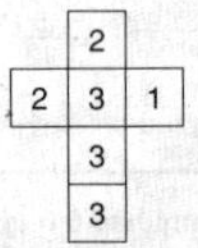

1. Predict which number you will land on most often. Explain.

   3; there are more chances to land on 3

2. Predict which number you will land on least often. Explain.

   1; there is only 1 chance to land on 1

Actual events from an experiment may or may not match your prediction. The chart shows the outcomes from 100 trials.

Outcome	1	2	3
Toss	28	39	33

Refer to the table for Exercises 3 and 4. Answer each question.

3. Did your prediction for landing on 1 match the outcome in the table?

   likely answer: no

4. Did your prediction for landing on 3 match the outcome in the table?

   likely answer: no

17
Holt Mathematics

Puzzles, Twisters, & Teasers — 11-2
All Bark, No Bite!

Find each experimental probability. Write your answer as a fraction, a decimal, and a percent. Then match the letters to the answers to solve the riddle.

1. Nick hits a target 7 out of 20 times. What is the experimental probability that he will hit the target on his next try?

   $\frac{7}{20}$; 0.35; 35%
   N

2. Melanie makes 12 out of 16 foul shots. What is the experimental probability that she will miss her next foul shot?

   $\frac{1}{4}$; 0.25; 25%
   B

3. A pollster surveys 80 people to determine whether they will vote for Seth Gleason for mayor. 42 out of 80 say yes. What is the experimental probability that the next person surveyed will say he or she plans to vote for Seth Gleason?

   $\frac{21}{40}$; 0.525; 52.5%
   K

4. A police radar measured the speed of 40 cars. The radar shows that 14 of the 40 cars are above the speed limit. What is the experimental probability that the next car will be traveling at or below the speed limit?

   $\frac{13}{20}$; 0.65; 65%
   A

5. Rhonda watches students entering the school. She sees that 8 out of 25 are wearing boots. What is the experimental probability that the next student that Rhonda sees will be wearing boots?

   $\frac{8}{25}$; 0.32; 32%
   R

6. A quarterback completes 10 of 16 passes. What is the experimental probability that his next pass will be incomplete?

   $\frac{3}{8}$; 0.375; 37.5%
   I

Where does a dog keep its car?

IN A <u>B</u> <u>A</u> <u>R</u> <u>K</u> <u>I</u> <u>N</u> G LOT

   0.25 65% 0.32 $\frac{21}{40}$ $\frac{3}{8}$ 35%

18
Holt Mathematics

62

Holt Mathematics

Practice A
Make a List to Find Sample Spaces

1. Lindsay flips a coin and rolls a 1–6 number cube at the same time. What are the possible outcomes?

   (H, 1), (H, 2), (H, 3), (H, 4), (H, 5), (H, 6),

   (T, 1), (T, 2), (T, 3), (T, 4), (T, 5), (T, 6)

2. Jordan has a choice of wheat bread or rye bread and a choice of turkey, ham, or tuna for lunch. What are all the possible choices of sandwiches he can have?

   turkey on wheat, turkey on rye, ham on wheat,

   ham on rye, tuna on wheat, tuna on rye

3. Marisol has to decide whether to study Italian, French, or Spanish, and whether to take golf, tennis, or archery in gym class. What are the possible choices that Marisol has?

   Italian and golf, French and golf, Spanish and golf, Italian and tennis,

   French and tennis, Spanish and tennis, Italian and archery,

   French and archery, Spanish and archery

Choose the letter for the best answer.

4. Chad and Victoria are playing a game with a quarter and a spinner divided into sixths, numbered 1–6. Each player spins the spinner and tosses the coin. How many outcomes are possible in the game?

   A 2   **C** 10
   B 8   **(D)** 12

5. For a snack, Sophie can choose milk, apple juice, orange juice, or punch. To go with her drink, she can choose a chocolate cupcake, oatmeal cookie, or crackers. How large is the sample space?

   (F) 12   **H** 4
   G 7   **J** 3

6. Marva has a spinner divided into fourths and a 1–6 number cube. She spins the spinner and rolls the number cube. How many outcomes are possible in the game?

   A 4   **C** 10
   B 6   **(D)** 24

7. Larry has a choice of vanilla, chocolate, or strawberry ice cream. The choices of toppings are nuts, sprinkles, or coconut. How many one-topping sundaes can he make?

   F 3   **(H)** 9
   G 6   **J** 12

19

Holt Mathematics

Practice B
Make a List to Find Sample Spaces

1. Marcus spins the spinner at the right and flips a dime at the same time. What are the possible outcomes? How many outcomes are in the sample space?

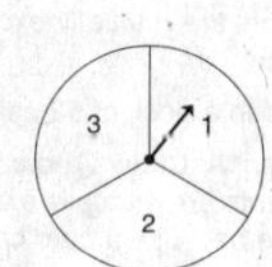

   (1, H), (2, H), (3, H), (1, T), (2, T),

   (3, T); 6 possible outcomes

2. For lunch, Britney has a choice of a hot dog, a hamburger, or pizza and a choice of an apple, a pear, or grapes. What are all the possible choices of lunch she can have? How many outcomes are in the sample space?

   hot dog with an apple, hot dog with a pear, hot dog with grapes,

   hamburger with an apple, hamburger with a pear,

   hamburger with grapes, pizza with an apple, pizza with a pear,

   pizza with grapes; 9 possible outcomes

3. Susan and Ryan are playing a game that involves spinning the spinner at the right and flipping a penny. How many outcomes are possible in the game?

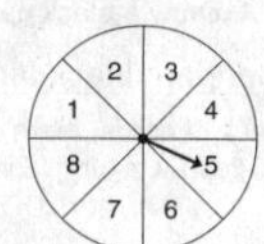

   16 outcomes

4. An Italian restaurant offers small, medium, and large calzones. The choices of fillings are cheese, sausage, spinach, or vegetable. How many different calzones can you order?

   12 different calzones

5. There are 5 ways to go from Town X to Town Y. There are 3 ways to go from Town Y to Town Z. How many different ways are there to go from Town X to Town Z, passing through Town Y?

   15 ways

6. Rasheed has tan pants, black pants, gray pants, and blue pants. He has a brown sweater and a white sweater. How many different ways can he wear a sweater and pants together?

   8 ways

20

Holt Mathematics

Practice C
Make a List to Find Sample Spaces

1. Joanna spins the spinners at the right at the same time. What are the possible outcomes? How many outcomes are in the sample space?

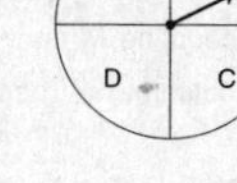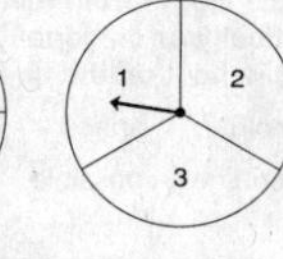

   (A, 1), (A, 2), (A, 3), (B, 1),

   (B, 2), (B, 3), (C, 1), (C, 2),

   (C, 3), (D, 1), (D, 2), (D, 3);

   12 possible outcomes

2. For breakfast, Armando has a choice of pancakes, eggs, or cereal and a choice of milk, hot cocoa, or juice. What are all the possible choices of breakfast he can have? How many outcomes are in the sample space?

   pancakes and milk, pancakes and hot cocoa, pancakes and juice,

   eggs and milk, eggs and hot cocoa, eggs and juice, cereal and milk,

   cereal and hot cocoa, cereal and juice; 9 possible outcomes

3. Shannon and Tyler are playing a game that involves spinning the spinner shown at the right and tossing a 1–6 number cube. How many outcomes are possible in the game?

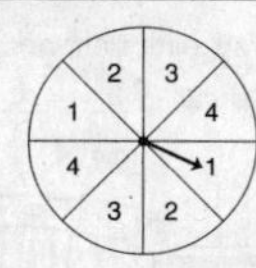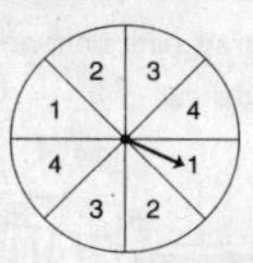

   24 outcomes

4. If you flip a penny, toss a 1–6 number cube, and flip a quarter, how many outcomes are possible?   24 outcomes

5. A Chinese restaurant has a special on Friday nights. For $20, you can choose one dish from 6 choices in column A and one dish from 5 choices in Column B. In addition, you can choose egg drop or wonton soup. How many different specials can you order?   60 specials

6. Lisa has a beige skirt, a black skirt, and a denim skirt. She has a red sweater and a white sweater, and she has a white blouse, a blue blouse, and a green blouse. How many different ways can she wear a skirt, sweater, and blouse together?   18 ways

21

Holt Mathematics

Reteach
Make a List to Find Sample Spaces

The set of all possible outcomes to an experiment is called the **sample space.**

A coin is tossed, and a number cube is rolled. What are all the possible outcomes? How many outcomes are in the sample space? There are two ways to show the sample space for an experiment. You can make a list, or you can make a tree diagram.

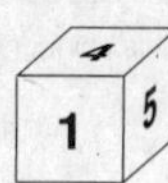

Make a list.

1. The possible outcomes for tossing a coin are __heads__ (H) and __tails__ (T).

2. The possible outcomes for rolling a number cube are __1__, __2__, __3__, __4__, __5__, and __6__.

3. The sample space is (H, 1), (H, 2), (H, 3), (H, 4), (H, 5), (H, 6), (T, 1), (T, 2), (T, 3), (T, 4), (T, 5), (T, 6). There are __12__ possible outcomes in the sample space.

Make a tree diagram.

4. 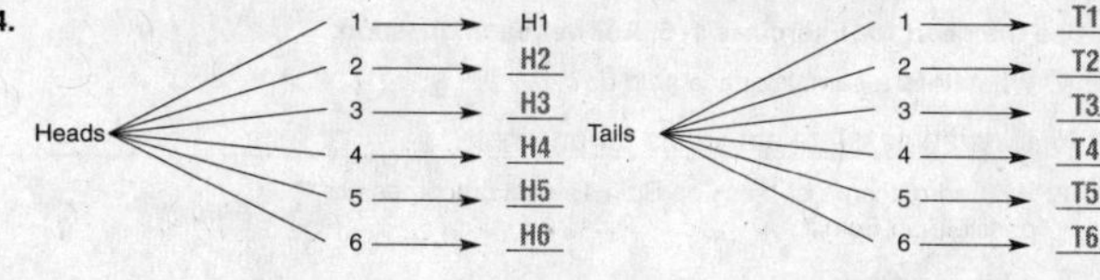

You can also find the number of possible outcomes by using the Fundamental Counting Principle.

Multiply the possible outcomes of each event.

5. 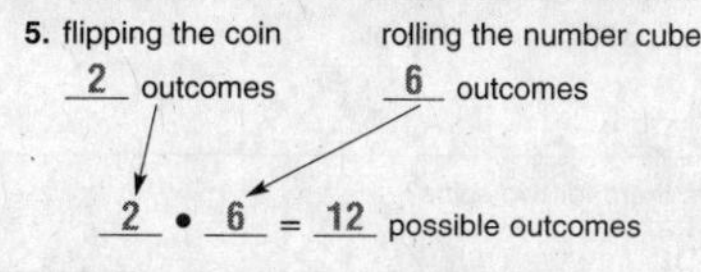

   flipping the coin   rolling the number cube
   __2__ outcomes   __6__ outcomes

   __2__ • __6__ = __12__ possible outcomes

22

Holt Mathematics

63

Holt Mathematics

Challenge
Mutually Exclusive Cards?

Two events are *mutually exclusive* if they cannot occur at the same time.

Example: In a deck of 52 cards, there are 26 red and 26 black cards.

Event A: Draw a red card. *Event B*: Draw a black card. The events are mutually exclusive. They cannot occur at the same time because a card cannot be red and black.

Event A: Draw a red card. *Event B*: Draw a number card. The events are not mutually exclusive. They can occur at the same time since a number card may be red.

For each pair of events, tell whether the two events are mutually exclusive for a single experiment. If they are not, explain why.

1. In a deck of cards, there are 40 number cards and 12 face cards. *Event A*: Draw a number card. *Event B*: Draw a face card.

   yes

2. In a deck of cards, there are 26 black cards and 12 face cards. *Event A*: Draw a black card. *Event B*: Draw a face card.

   no; there are black face cards

3. In a deck of cards, there are 4 kings and 4 queens. *Event A*: Draw a king. *Event B*: draw a queen.

   yes

4. In a deck of cards, there are 12 diamonds and 12 hearts. *Event A*: Draw a diamond. *Event B*: Draw a heart.

   yes

5. In a deck of cards, there are 12 diamonds and 12 face cards. *Event A*: Draw a diamond. *Event B*: Draw a face card.

   no; you could draw the King of diamonds

6. In a deck of cards, there are 40 number cards and 4 jacks. *Event A*: Draw a number card. *Event B*: Draw a jack.

   yes

7. In a deck of cards, there are 4 tens and 13 clubs. *Event A*: Draw a ten. *Event B*: Draw a club.

   no; you could draw the ten of clubs

23
Holt Mathematics

Problem Solving
Make a List to Find Sample Spaces

Write the correct answer.

Benny's Bagels	
Bagels	**Toppings**
Plain	Cream cheese
Poppy	Honey
Raisin	Butter
Sesame	Jam
Egg	

1. If you order one topping, how many different choices of bagel and toppings can you order?

   20 choices

2. Santana only likes cream cheese or jam on his bagel. How many choices does he have for a one-topping bagel?

   10 choices

3. Yesterday, Benny ran out of raisin bagels. How many choices of a bagel and one topping were there?

   16 choices

4. Today, Benny has all 5 types of bagels but runs out of honey. How many choices of a bagel with one topping can you order?

   15 choices

Choose the letter for the best answer.

5. The mall movie multiplex is showing 12 movies. Each movie is shown at five different times during the day. How many choices of movies and showtimes does Reggie have?

   A 5   C 17
   B 12   (D) 60

6. At Hi-Top Ski Resort, there are three chair lifts to the top of the mountain. There are six ski trails to the bottom of the mountain. How many possible choices of lifts and trails do the skiers have?

   F 9   H 81
   (G) 18   J 2

7. In a Little League game, Geri can bat first, second, or third. When at bat, she could strike out, walk, or get a hit. How many outcomes are in the sample space for these events?

   A 3
   B 6
   (C) 9
   D 18

8. Ty is flipping a coin. He has decided that if he flips the same result twice in a row, he will do his homework. If he flips 2 different results, then he will go jogging. How likely is it that he will study?

   (F) as likely as not
   G likely
   H unlikely
   J certain

24
Holt Mathematics

Reading Strategies
Read a Chart

To measure the probability of an outcome, you must find all the possible outcomes to the experiment. A **sample space** lists all possible outcomes to an experiment.

This spinner can land on white or black.

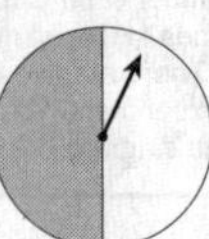

Each line of this chart lists a possible outcome from two spins.

Spin 1	Spin 2
Black	Black
White	Black
Black	White
White	White

Use the chart for Exercises 1–5. Answer each question.

1. What does a sample space help do?

   Lists all possible outcomes for an event.

2. With an outcome of black on Spin 1, what outcomes are possible on Spin 2?

   white or black

3. With an outcome of black on Spin 1, one outcome for Spin 2 is black. This is shown as "black–black." How would you write the other possible outcome after Spin 2?

   black–white

4. With an outcome of white on Spin 1, what are the possible outcomes for Spin 2?

   white or black

5. How many possible outcomes are there for two spins?

   4

25
Holt Mathematics

Puzzles, Twisters & Teasers
Sealed With a Fish

Circle words from the list in the word search (horizontally, vertically or diagonally). Find a word that answers the riddle and write it on the line.

sample	space	counting	principle	fundamental
determine	possible	outcome	tree	diagram

```
D E T E R M I N E W E R O
I P R I N C I P L E Y P U
A Q N Z T C V B N M K L T
G W E W E R T Y U I O S C
R A S D F G E J K L P E O
A Q A Z X S W E D C V A M
M F U N D A M E N T A L E
N Z X C V B G T S H U J I
C O U N T I N G I P O L T
R C V B G T Y H N J A I K
P O S S I B L E N J U C D
S A M P L E V G T E D F E
```

What Is gray, eats fish, and lives in Washington, D.C.?

The presidential   S E A L .

26
Holt Mathematics

Practice A
Theoretical Probability

Tina has 3 quarters, 1 dime, and 6 nickels in her pocket. Find the probability of randomly drawing each of the following coins. Write your answer as a fraction, as a decimal, and as a percent.

		Fraction	Decimal	Percent
1.	quarter	$\frac{3}{10}$	0.3	30%
2.	dime	$\frac{1}{10}$	0.1	10%
3.	nickel	$\frac{6}{10} = \frac{3}{5}$	0.6	60%

Find the probability of each event. Write your answer as a fraction, as a decimal, and as a percent. Round to the nearest tenth of a percent.

4. randomly choosing a red card in a game that has 10 red, 10 blue, 10 green, 10 yellow cards, and 10 orange cards

$\frac{1}{5}$; 0.2; 20%

5. tossing two fair coins and having both land tails up

$\frac{1}{4}$; 0.25; 25%

6. randomly drawing 1 of the 4 S's from a bag of 100 Scrabble tiles

$\frac{1}{25}$; 0.04; 4%

7. rolling a number greater than 4 on a fair number cube

$\frac{1}{3}$; 0.333; 33.3%

A game has 12 blue disks, 10 red disks, and 8 black disks. Find the probability of each event when a disk is chosen at random.

8. red $\frac{1}{3}$

9. black $\frac{4}{15}$

10. blue $\frac{2}{5}$

11. not red or blue $\frac{4}{15}$

27
Holt Mathematics

Practice B
Theoretical Probability

Find the probability of each event. Write your answer as a fraction, as a decimal, and as a percent. Round to the nearest tenth of a percent.

1. randomly choosing a white counter from a bag of 12 red counters, 12 white counters, 12 green counters, and 12 blue counters

$\frac{1}{4}$; 0.25; 25%

2. tossing two fair coins and having one land on tails and one land on heads

$\frac{1}{2}$; 0.5; 50%

3. rolling a number greater than 1 on a fair number cube

$\frac{5}{6}$; 0.833; 83.3%

4. randomly drawing an orange disk from a bag of 14 black disks, 4 blue disks and 12 orange disks

$\frac{2}{5}$; 0.4; 40%

5. randomly drawing 1 of the 6 R's from a bag of 100 Scrabble tiles

$\frac{3}{50}$; 0.06; 6%

6. spinning a number less than 7 on a fair spinner with 8 equal sections labeled 1-8

$\frac{3}{4}$; 0.75; 75%

A set of cards has 20 cards with stars, 10 cards with squares, and 15 cards with circles. Find the probability of each event when a card is chosen at random.

7. square $\frac{2}{9}$

8. circle $\frac{1}{3}$

9. star or circle $\frac{7}{9}$

10. not circle or square $\frac{4}{9}$

There are 14 girls and 18 boys in Ms. Wiley's class. Ms. Wiley randomly selects one student to solve a problem. Find the probability of each event.

11. selecting a boy $\frac{9}{16}$

12. selecting a girl $\frac{7}{16}$

28
Holt Mathematics

Practice C
Theoretical Probability

Find the probability of each event. Write your answer as a fraction, as a decimal, and as a percent. Round to the nearest tenth of a percent.

1. tossing three fair quarters and having all three of them land on heads

$\frac{1}{8}$; 0.125; 12.5%

2. randomly choosing a classical CD from a collection of CDs consisting of 35 jazz CDs, 20 classical CDs, 25 rock CDs, and 5 country music CDs

$\frac{4}{17}$; 0.235; 23.5%

3. randomly choosing a card with an even number from a shuffled deck of 52 cards with four 13-card suits (diamonds, hearts, clubs and spades), each of which has 9 number cards labeled 2-10 and 4 other cards

$\frac{5}{13}$; 0.385; 38.5%

4. randomly drawing a vowel from a bag of 100 Scrabble® tiles that has 12 E's, 9 I's, 8 O's, 4 U's and 2 Y's.

$\frac{7}{20}$; 0.35; 35%

There are 15 girls and 9 boys in Anne's yoga class. One of them is randomly selected to demonstrate a yoga position. Find the probability of each event.

5. selecting a boy $\frac{3}{8}$

6. selecting a girl $\frac{5}{8}$

Find the probability of each event when two 1-6 number cubes are rolled.

7. P(total of 5) $\frac{1}{9}$

8. P(total of 10) $\frac{1}{12}$

9. P(total ≥7) $\frac{7}{12}$

10. P(total < 2) 0

29
Holt Mathematics

Reteach
Theoretical Probability

The **theoretical probability** of an event is found by comparing the number of ways an event can occur to the total number of equally likely outcomes.

$$\text{theoretical probability} = \frac{\text{number of ways the event can occur}}{\text{total number of equally likely outcomes}}$$

One of the games at a carnival is the Wheel of Letters. Find the probability that the wheel will stop on each letter. Write your answer as a fraction, as a decimal, and as a percent.

1. The spinner has $\underline{10}$ equal sections. Each section is an equally likely outcome.

2. There is $\underline{1}$ section marked A.

3. There are $\underline{3}$ sections marked B.

4. There are $\underline{4}$ sections marked C.

5. There are $\underline{2}$ sections marked D.

6. $P(B) = \frac{3}{10} = 0.3 = \underline{30\%}$

7. $P(A) = \frac{1}{10} = 0.1 = \underline{10\%}$

8. $P(C) = \frac{4}{10} = \frac{2}{5} = 0.4 = \underline{40\%}$

9. $P(D) = \frac{2}{10} = \frac{1}{5} = 0.2 = \underline{20\%}$

There are 11 pennies and 9 dimes in a bag. Find the probability of each event. Write your answer as a fraction, as a decimal, and as a percent.

10. Find the probability that a dime will be drawn from the bag.

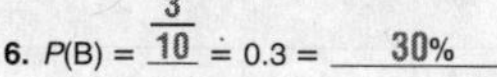

P(dime) $= \frac{9}{20} = 0.45 = \underline{45\%}$

11. P(penny) $= \frac{11}{20} = 0.55 = \underline{55\%}$

There are 6 yellow cards, 4 blue cards, and 10 green cards in a bag. A card is chosen at random. Find the probability of each event.

12. yellow $\frac{3}{10}$

13. blue $\frac{1}{5}$

14. green $\frac{1}{2}$

15. blue or green $\frac{7}{10}$

30
Holt Mathematics

Challenge
Spinner Sums

Suppose these three spinners are each spun once.

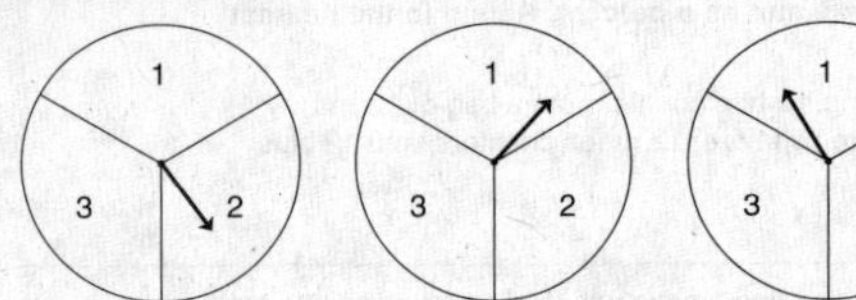

1. Complete the organized list so that it shows every sum that is possible if all three spinners are spun once.

First Spinner Shows 1	1 + 1 + 1 = 3 1 + 1 + 2 = 4 1 + 1 + 3 = 5	1 + 2 + 1 = 4 1 + 2 + 2 = 5 1 + 2 + 3 = 6	1 + 3 + 1 = 5 1 + 3 + 2 = 6 1 + 3 + 3 = 7
First Spinner Shows 2	2 + 1 + 1 = 4 2 + 1 + 2 = 5 2 + 1 + 3 = 6	2 + 2 + 1 = 5 2 + 2 + 2 = 6 2 + 2 + 3 = 7	2 + 3 + 1 = 6 2 + 3 + 2 = 7 2 + 3 + 3 = 8
First Spinner Shows 3	3 + 1 + 1 = 5 3 + 1 + 2 = 6 3 + 1 + 3 = 7	3 + 2 + 1 = 6 3 + 2 + 2 = 7 3 + 2 + 3 = 8	3 + 3 + 1 = 7 3 + 3 + 2 = 8 3 + 3 + 3 = 9

2. Complete the table using the chart of sums in Exercise 1.

Sum of Spinners	3	4	5	6	7	8	9
Number of Ways to Get the Sum	1	3	6	7	6	3	1

3. How many possible outcomes are there are if you spin the three spinners? __27__

Using the table in Exercise 2, give the probability of getting each sum. Write the probability to the nearest tenth of a percent.

4. P(sum of 5)

__22.2%__

5. P(sum > 7)

__14.8%__

6. P(sum < 8)

__85.2%__

31   **Holt Mathematics**

Problem Solving
Theoretical Probability

Write the correct answer in simplest form.

The table lists the pieces used in the game of chess.

Chess Set		
Type	White	Black
Pawn	8	8
Rook	2	2
Knight	2	2
Bishop	2	2
Queen	1	1
King	1	1

1. If you randomly choose a piece of the chess set, what is the probability that it is a pawn? Write your answer as a decimal.

__0.5__

2. If you randomly choose a piece of the chess set, what is the probability that it is a white pawn? Write your answer as a decimal.

__0.25__

3. If you randomly choose a piece of the chess set, what is the probability that it is a rook, knight, or bishop? Write your answer as a fraction.
$\frac{3}{8}$

4. If you randomly choose a piece of the chess set, what is the probability that it is a queen? Write your answer as a fraction.
$\frac{1}{16}$

Choose the letter for the best answer.

5. Mr. Rose draws names to see who will give the first book report. There are 10 boys and 14 girls in his class. What is the probability that he will draw a girl's name?

   A $\frac{2}{5}$ Ⓒ $\frac{7}{12}$

   B $\frac{5}{12}$ D $\frac{5}{7}$

6. Eight students will give reports on novels, 9 will report on biographies, and 7 will report on history books. What is the probability that the first report will be a novel?

   F $\frac{3}{8}$ H $\frac{8}{17}$

   Ⓖ $\frac{1}{3}$ J $\frac{1}{2}$

7. Stanley is reading a 224-page book. There are illustrations on 14 pages. If Stanley opens the book at random, what is the probability that the page will have an illustration?

   A 0.0004   C 0.0714

   Ⓑ 0.0625   D 0.9375

8. In Congress, each of the 50 states is represented by 2 senators. If you choose a senator randomly, what is the probability that you will choose a senator that represents Virginia?

   F 25%   Ⓗ 2%

   G 10%   J 1%

32   **Holt Mathematics**

Reading Strategies
Interpret Information

When you are conducting an experiment, a **favorable outcome** is an outcome you are looking for.

Theoretical Probability

$$\text{probability(event)} = \frac{\text{number of ways an outcome can occur}}{\text{total number of possible outcomes}}$$

When you toss a number cube, there are 6 possible outcomes. There is only one way to get a 4 on the cube, so there is one favorable outcome for rolling a 4.

Read: The probability of rolling a 4 is 1 out of 6.

Write: $P(\text{rolling a 4}) = \frac{1}{6}$

Answer each question.

1. What is the probability of rolling a two on the cube?

   1 out of 6 or $\frac{1}{6}$

2. What are the even numbers on the cube?

   2, 4, and 6

3. What is the probability of rolling an even number?

   3 out of 6 or $\frac{3}{6}$ or $\frac{1}{2}$

4. What are the odd numbers on the cube?

   1, 3, and 5

5. What is the probability of rolling an odd number.

   3 out of 6 or $\frac{3}{6}$ or $\frac{1}{2}$

If events have the same chance of happening, they are **equally likely** to occur.

6. Is the probability of rolling a 5 or an even number equally likely?

   no

7. Is the probability of rolling an odd or an even number equally likely?

   yes

33   **Holt Mathematics**

Puzzles, Twisters & Teasers
Probably Problems!

Across

1. ____ probability is used to calculate the probability of an event when all outcomes are equally likely.
3. You can write a ____ as a decimal or a fraction.
5. If each possible outcome of an experiment is equally likely, the experiment is said to be ____.
7. You can write a ____ as a fraction or a percent.
8. A ____ outcome is one that you are looking for when you conduct an experiment.

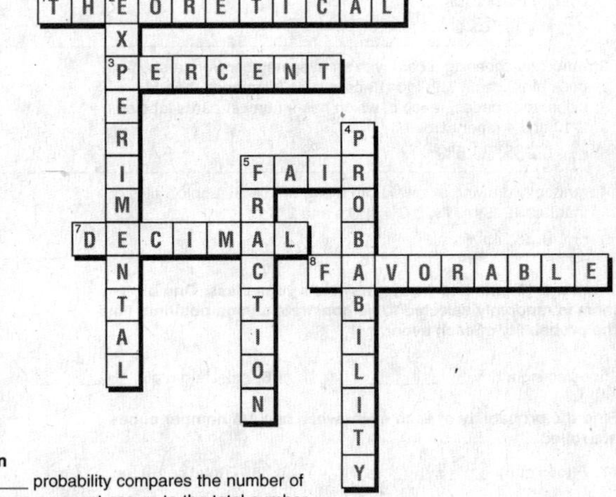

Down

2. ____ probability compares the number of times an event occurs to the total number of trials.
4. You can write ____ as a fraction, a decimal, or a percent.
5. You can write a ____ as a decimal or as a percent.

34   **Holt Mathematics**

Holt Mathematics

Practice A
Probability of Independent and Dependent Events

Decide if each set of events is independent or dependent. Explain your answer.

1. A student spins a spinner and rolls a number cube.

   Independent; Spinning a spinner does not affect
   the outcome of rolling a number cube.

2. A student picks a raffle ticket from a box and then picks a second raffle ticket without replacing the first raffle ticket.

   Dependent; There are fewer raffle tickets in the
   box for the second ticket picked.

Find the probability of each set of independent events. Choose the letter for the best answer.

3. drawing a black checker from a bag of 6 black checkers and 4 red checkers, replacing it, and drawing another black checker

   A $\frac{2}{3}$ C $\frac{2}{5}$
   B $\frac{9}{25}$ D $\frac{3}{5}$

4. rolling a six on the first roll of a 1–6 number cube and rolling an odd number on the second roll of the same cube

   F $\frac{1}{12}$ H $\frac{1}{6}$
   G $\frac{1}{8}$ J $\frac{1}{2}$

5. flipping a tail on a coin and spinning a 5 on a spinner with sections of equal area numbered 1–5

   A $\frac{1}{2}$ C $\frac{1}{7}$
   B $\frac{1}{5}$ **D** $\frac{1}{10}$

6. drawing a 1, 2, or 3 from 9 cards numbered 1–9, replacing the card, and drawing a 7, 8, or 9

   F $\frac{1}{3}$ **H** $\frac{1}{9}$
   G $\frac{3}{8}$ J $\frac{1}{12}$

Solve.

7. There are 4 black marbles and 2 white marbles in a bag. What is the probability of choosing a black marble, not replacing it, then choosing a white marble?

   $\frac{4}{15}$

 Holt Mathematics

Practice B
Probability of Independent and Dependent Events

Decide if each set of events is independent or dependent. Explain your answer.

1. A student spins a spinner and chooses a Scrabble® tile

   Independent; Spinning a spinner does not affect
   the outcome of choosing a Scrabble® tile.

2. A boy chooses a sock from a drawer of socks, then chooses a second sock without replacing the first.

   Dependent; There are fewer socks in the drawer
   for the second sock picked.

3. A student picks a raffle ticket from a box, replaces the ticket, then picks a second raffle ticket.

   Independent; There are the same number of raffle
   tickets in the box for the second ticket picked.

Find the probability of each set of independent events.

4. drawing a red checker from a bag of 9 black checkers and 6 red checkers, replacing it, and drawing another red checker

   $\frac{4}{25}$

5. drawing a black checker from a bag of 9 black checkers and 6 red checkers, replacing it, and drawing a red checker

   $\frac{6}{25}$

6. rolling a 1, 2, or 3 on the first roll of a 1–6 number cube and rolling a 4, 5, or 6 on the second roll of the same cube

   $\frac{1}{4}$

Solve.

7. Randy has 4 pennies, 2 nickels, and 3 dimes in his pocket. If he randomly chooses 2 coins, what is the probability that both are dimes?

   $\frac{1}{12}$

 Holt Mathematics

Practice C
Probability of Independent and Dependent Events

Decide if each set of events is independent or dependent. Explain your answer.

1. A student spins an even number on a spinner and then spins another even number on the second spin.

   Independent; The first spin does not affect the
   outcome of the second spin.

2. A student guesses on two multiple choice questions.

   Independent; Each question is a separate event.

Find the probability of each set of independent events.

3. drawing a brown sock from a drawer of 4 brown socks, 10 black socks, and 6 gray socks, replacing it, then drawing a gray sock

   $\frac{3}{50}$

4. drawing a pair of black socks from a drawer of 4 brown socks, 10 black socks, and 6 gray socks

   $\frac{9}{38}$

Solve.

5. Calista has 4 one-dollar bills, 2 five-dollar bills, and 3 ten-dollar bills in her wallet. If she randomly chooses 2 bills from her wallet, what is the probability that both are five dollar bills?

   $\frac{1}{36}$

6. There are 10 true/false questions on a test. You do not know the answer to 4 of the questions, so you guess. What is the probability that you will get all 4 answers right?

   $\frac{1}{16}$

7. There are 12 students in the After-School Club. Three names are being drawn at random to help with the school carnival. What is the probability that Jess, Sandy, and Phil will be chosen?

   $\frac{1}{220}$

 Holt Mathematics

Reteach
Probability of Independent and Dependent Events

Events are *independent* when the outcome of one event has no effect on the outcome of a second event. Rolling a number cube and flipping a coin are **independent events**.

Find the probability of rolling a 4 and flipping heads.

1. There are __6__ outcomes for the number cube and __2__ outcomes for the coin.

2. Using the Fundamental Counting Principle, there are __6__ × __2__, or __12__, possible outcomes of rolling a number cube and flipping a coin.

3. Make a list of the possible outcomes:

   (1, H), (2, H), (3, H), (4, H), (5, H), (6, H), (1, T), (2, T),
   (3, T), (4, T), (5, T), (6, T)

   How many possible ways are there of rolling a 4 and flipping heads? __1__

4. $P(4 \text{ and heads}) = \dfrac{\text{number of ways 4 and heads can occur}}{\text{number of possible outcomes}} = \dfrac{1}{12}$

Events are *dependent* when the outcome of one event does have an effect on the outcome of the next event. Drawing two marbles in a row from a bag without replacing the first marble are **dependent events**.

A bag contains 3 blue and 5 red marbles. Find the probability of drawing 2 blue marbles in a row without replacing the first marble.

5. The total number of marbles in the bag is __8__. There are __3__ blue marbles in the bag. $P(\text{blue marble on first draw}) = \dfrac{3}{8}$

6. There are now __7__ marbles in the bag. If you drew a blue marble on the first draw, there are __2__ blue marbles left in the bag.

   $P(\text{blue marble on second draw}) = \dfrac{2}{7}$

7. $P(\text{blue, blue}) = P(\text{blue on 1st draw}) \times P(\text{blue on 2nd draw}) = \dfrac{3}{8} \times \dfrac{2}{7} = \dfrac{3}{28}$

 Holt Mathematics

Holt Mathematics

Challenge
Pascal's Triangle

A special pattern, called Pascal's Triangle, can be used to find some probabilities. The triangle is called Pascal's Triangle because Pascal was one of the first mathematicians to formally study probability.

```
Row
0                    1
1                  1   1
2                1   2   1
3              1   3   3   1
4            1   4   6   4   1
5          1   5  10  10   5   1
```

Suppose 3 coins are flipped.

1. List all the possible outcomes.

   HHH, HHT, HTH, THH, HTT, THT, TTH, TTT

2. How many outcomes show: all heads? 2 heads and 1 tail? 2 tails and 1 head? 3 tails?

   1, 3, 3, 1

3. How do the outcomes compare to row 3 of Pascal's Triangle?

   They are the same.

Find the probability of each event when flipping 3 coins.

4. 3 heads? $\frac{1}{8}$

5. 2 heads? $\frac{3}{8}$

6. 1 head? $\frac{3}{8}$

7. 0 heads? $\frac{1}{8}$

Use Pascal's Triangle to find the following probabilities if 4 coins are flipped.

8. 4 tails $\frac{1}{16}$

9. 3 tails $\frac{4}{16}$, or $\frac{1}{4}$

10. 2 tails $\frac{6}{16}$, or $\frac{3}{8}$

11. 1 tail $\frac{4}{16}$, or $\frac{1}{4}$

12. 0 tails $\frac{1}{16}$

13. 2 heads $\frac{6}{16}$, or $\frac{3}{8}$

Holt Mathematics

Problem Solving
Probability of Independent and Dependent Events

Write the correct answer.

1. Li rolls a pair of number cubes twice. On both rolls, the sum is 7. Are the rolls dependent or independent events?

   independent events

2. Nine boys and 12 girls want to play soccer. Teams are formed by selecting one player at a time. Is the probability of selecting a boy after a girl is selected a dependent or an independent event?

   dependent event

3. Gregg has 12 cards. Half are black, and half are red. He picks two cards out of the deck. What is the probability that both cards are red?

   $\frac{5}{22}$

4. In basketball, Alan makes 1 out of every 4 free throws he attempts. What is the probability that Alan will make his next 3 free throws?

   $\frac{1}{64}$

Choose the letter for the best answer.

5. There are 8 blue marbles and 7 red marbles in a bag. Julie pulls two marbles at random from the bag first. What is the probability that she first pulls a blue marble and then a red marble?

   A $\frac{8}{15}$　　C $\frac{4}{7}$

   B $\frac{4}{15}$　　D $\frac{1}{2}$

6. You roll a 1–6 number cube twice. What is the probability that you roll a 3 on the first roll and a 6 on the second roll?

   F $\frac{1}{36}$　　H $\frac{1}{6}$

   G $\frac{1}{9}$　　J $\frac{1}{2}$

7. Andrew has $2.00 in quarters in his pocket, including three state quarters. He takes two quarters out of his pocket. What is the probability that they are **not** state quarters?

   A $\frac{3}{8}$　　C $\frac{3}{14}$

   B $\frac{5}{8}$　　D $\frac{5}{14}$

8. Jamie has 3 raffle tickets. One hundred tickets were sold. Her name was not drawn for the first prize. What is the probability that her name will be drawn for the second prize?

   F $\frac{1}{3}$　　H $\frac{1}{33}$

   G $\frac{3}{100}$　　J $\frac{2}{99}$

Holt Mathematics

Reading Strategies
Focus on Vocabulary

When the outcome of one event does not affect the probability of the outcome of another, they are called **independent events**.

If you flip a coin once, it can land on head or tails.

$P(\text{landing on tails}) = \frac{1}{2}$

If you flip a coin a second time, $P(\text{landing on tails})$ is still $\frac{1}{2}$.

When the outcome of one event affects the probability of the outcome of another event, the events are called **dependent events**.

There are 52 cards in a deck, and 4 aces in the deck.

$P(\text{drawing an ace}) = \frac{4}{52}$

If an ace is drawn and not replaced, the probability of drawing an ace on the second draw drops to 3 out of 51.

$P(\text{drawing another ace}) = \frac{3}{51}$

Write *dependent events* or *independent events* to complete each sentence.

1. You roll a cube. Then you roll the cube a second time.

   independent events

2. You take a coin from the jar and do not replace it. Then you take another coin from the jar.

   dependent events

3. One person in the class is chosen to be first in line. Another person is chosen to turn off the lights.

   dependent events

4. You have a bag with red, white, and blue counters. Explain how to conduct an experiment pulling counters from the bag so you will have independent events.

   Every time you pull a counter from the bag you replace it before the next trial.

5. Explain how to change the experiment so the events are dependent.

   When you pull a counter from the bag, do not replace it before the next trial.

Holt Mathematics

Puzzles, Twisters & Teasers
Furry Math!

Decide whether each event is dependent or independent. Circle the letter above your answer. Match the letters with the number of the problems to solve the riddle.

1. tossing a coin twice
   Q　　(T)
   dependent　　independent

2. spinning a spinner five times
   M　　(H)
   dependent　　independent

3. pulling two socks from a drawer at the same time
   (E)　　P
   dependent　　independent

4. drawing two marbles out of a bag at the same time
   (O)　　D
   dependent　　independent

5. throwing a pair of dice ten times
   X　　(U)
   dependent　　independent

6. drawing two names out of a hat without replacement
   (T)　　B
   dependent　　independent

7. picking cards from a deck without replacement
   (S)　　V
   dependent　　independent

8. spinning two different spinners two times
   K　　(I)
   dependent　　independent

9. throwing three coins three times
   A　　(D)
   dependent　　independent

10. throwing one die five times
    U　　(E)
    dependent　　independent

Which side of a rabbit has the most fur?

T H E O U T S I D E
1 2 3 4 5 6 7 8 9 10

Holt Mathematics

Holt Mathematics

Practice A
Combinations

Solve each problem in Column A. Draw a line to the correct answer in Column B.

Column A	Column B

1. If you have watermelon, cantaloupe and honeydew, how many combinations of two fruits are there?

   A. 20

2. How many three-letter combinations are possible using the letters D, E, F, G, and H?

   B. 15

3. How many different two-person debating teams can be chosen from six students?

   C. 21

4. On Mondays at Pizza Pan, you can choose two toppings at no extra cost for your pizza. The toppings are mushrooms, sausage, peppers, and extra cheese. How many different pizzas could you order?

   D. 10

5. How many different three-person relay teams are possible with four people?

   E. 3

6. The students in Mr. Trumbull's English class need to read 2 books from a list of 7 books. How many different combinations of books are possible?

   F. 6

7. Erin has 6 colors of ribbon: red, white, gold, green, blue, and purple. She makes bows of out 3 different colors of ribbon. How many different combinations of colors can she choose?

   G. 4

43

Practice B
Combinations

1. A chef has some broccoli, cauliflower, carrots, and squash to make a vegetarian dish. List the possible combinations if he uses only 3 vegetables in the dish.

broccoli, cauliflower, carrots; broccoli, cauliflower, squash; broccoli, carrots, squash; cauliflower, carrots, squash

2. Lauren, Manuel, Nick, Opal, and Pat are forming groups of two to work on a drama production. List the different combinations of students that are possible using the first initial of each name.

LM, LN, LO, LP, MN, MO, MP, NO, NP, OP

3. Keiko has seven colors of lanyard. She uses three different colors to make a key chain. How many different combinations can she choose?

35 combinations

4. On Sundays at Ice Cream Heaven, you can choose two free toppings for your sundae. The toppings are nuts, hot fudge, caramel, and sprinkles. How many different combinations of toppings can you order?

6 combinations

5. How many different three-person relay teams can be chosen from six students?

20 teams

6. The students in Mrs. Mandel's class need to choose two class representatives from six nominated students. How many different combinations of class representatives are possible?

15 combinations

7. There are four varieties of muffins available at the Coffee Shop. How many different ways can you choose three different muffins?

4 ways

8. How many two-person carpools are possible with seven people?

21 carpools

44

Practice C
Combinations

1. If you have tomatoes, red peppers, onions, green peppers, and mushrooms, how many combinations of three vegetables are there?

10 combinations

2. How many three-letter combinations are possible from P, Q, R, S, T, and U?

20 combinations

3. Kim has seven colors of hair ribbons: red, white, pink, green, blue, black, and purple. She uses two different colors to make a bow. How many different combinations of colors can she choose?

21 combinations

4. Les, Dennis, Greg, and Philip are pairing up to play doubles tennis. In how many different ways can they pair up?

6 ways

5. You can choose two side dishes for your chicken dinner. The choices are mashed potatoes, roasted potatoes, squash, carrots, string beans, stuffing, broccoli, or corn. How many different side dish combinations could you order?

28 combinations

6. How many different five-person relay teams can be chosen from seven students?

21 relay teams

7. The students in Mr. Cohen's science class need to complete three experiments at home from a list of seven topics. How many different combinations of experiments are possible?

35 combinations

8. Lena, Drew, Michaela, Trey, Nancy, and Jonah are pairing up to play one-on-one basketball. In how many different ways can they pair up?

15 ways

Find the number of combinations.

9. 8 things taken 4 at a time

   70

10. 6 things taken 4 at a time

   15

11. 8 things taken 5 at a time

   56

12. 7 things taken 6 at a time

   7

13. 7 things taken 4 at a time

   35

14. 8 things taken 6 at a time

   28

45

Reteach
Combinations

A **combination** is a selection of objects in which the order is not important. You can make a list to find the number of combinations.

The school district is going to plant two types of trees around the playground. The landscaper has five kinds of trees to choose from.

Let the letters A, B, C, D, and E represent the different kinds of trees. You can make an organized list of all possible combinations.

Tree A with each other tree: AB   AC   AD   AE
Tree B with each other tree: BA   BC   BD   BE
Tree C with each other tree: CA   CB   CD   CE
Tree D with each other tree: DA   DB   DC   DE
Tree E with each other tree: EA   EB   EC   ED

1. How many groups of 2 trees are in the list?   20

2. AB is the same combination as _BA_   BD is the same combination as _DB_.

   AC is the same combination as _CA_   BE is the same combination as _EB_.

   AD is the same combination as _DA_   CD is the same combination as _DC_.

   AE is the same combination as _EA_   CE is the same combination as _EC_.

   BC is the same combination as _CB_   DE is the same combination as _ED_.

3. Cross out each of the duplications in the list above. How many are left?   10

4. Make an organized list of all possible two-letter combinations using the letters U, V, W, X, Y, and Z. Cross out each duplication.

UV, _UW_, _UX_, _UY_, _UZ_   VU, _VW_, _VX_, _VY_, _VZ_

WU, _WV_, _WX_, _WY_, _WZ_   XU, _XV_, _XW_, _XY_, _XZ_

YU, _YV_, _YW_, _YX_, _YZ_   ZU, _ZV_, _ZW_, _ZX_, _ZY_

There are _15_ combinations of 6 letters taken 2 at a time.

46

Challenge
You Can Count on It!

You can use the Fundamental Counting Principle to help you find a formula that gives the number of combinations that are possible when x objects are taken n objects at a time. Remember, the Fundamental Counting Principle states that if a first event can occur in a ways, and a second event can occur in b ways, then the two events can occur together in $a \cdot b$ ways.

Suppose there are six students. How many different teams of three students can be chosen?

1. Use the Fundamental Counting Principle to find how many ways three of the students can be chosen for a team.

Choices for the first student	Choices left for the second student	Choices left for the third student	Total number of ways
6	× 5	× 4	= 120 ways

2. Use the Fundamental Counting Principle to find how many ways any group of three students can be ordered.

Choices for the first student	Choices left for the second student	Choices left for the third student	Total number of orders
3	× 2	× 1	= 6 orders

3. The order in which the three students are chosen for the team does not matter, because changing the order does not change the team. So, to find the total number of teams that are possible, divide the number of ways three students can be chosen (Problem 1) by the total number of orders (Problem 2).

   120 ways ÷ 6 orders = 20 teams

Use the Fundamental Counting Principle and the formula, Number of Combinations = Number of Ways ÷ Number of Orders, to find the number of combinations.

4. 9 things taken 5 at a time

   126

5. 12 things taken 3 at a time

   220

6. 10 things taken 4 at a time

   210

7. 12 things taken 4 at a time

   495

47

Problem Solving
Combinations

Write the correct answer.

1. Six friends are going to play a ball game. Each team has 3 players. How many different team combinations are possible?

   20 combinations

2. Yung wants to visit 3 of the 5 Great Lakes this summer. How many different combinations of lakes are possible?

   10 combinations

3. There are 4 spots left for a school field trip. Roberta, Samuel, Thea, Ling, Jose, and Mark all want to go on the trip. How many different combinations are possible for the remaining 4 spots?

   15 combinations

4. At summer camp, the campers pick two activities for the first day of camp. They can choose from among hiking, canoeing, rock climbing, and bird watching. How many different combinations of activities are there?

   6 combinations

Choose the letter for the best answer.

5. Eight children are playing a trivia game. They want to make teams of 2 players each. How many different team combinations are possible?

   A 10 combinations
   B 21 combinations
   C 28 combinations
   D 56 combinations

6. At Washington Middle School, a student takes 4 core classes a day. There are 6 different core classes offered. How many different combinations of classes are there?

   F 10 combinations
   G 12 combinations
   H 15 combinations
   J 24 combinations

7. The class is drawing maps of the 7 continents to display in the school lobby. The main wall in the lobby has room for 3 maps. How many combinations of maps are possible for that location?

   A 10 combinations
   B 21 combinations
   C 30 combinations
   D 35 combinations

8. Ms. Henrie's literature class is voting on the 2 most influential people of the eighteenth century. The choices are 5 famous people. How many different combinations are possible?

   F 7 combinations
   G 8 combinations
   H 10 combinations
   J 15 combinations

48

Reading Strategies
Make an Organized List

Here are three different frozen yogurt flavors: vanilla, chocolate, strawberry. How many **combinations** of only two different flavors can you make?

Order does not matter. If you choose a scoop of vanilla first and a scoop of chocolate second, it is the same combination as choosing chocolate first and then vanilla.

Making a list can help you picture the possible combinations of two different flavors.

Scoop 1	Scoop 2
vanilla	chocolate
vanilla	strawberry
?	?

Use the chart to answer each question.

1. What flavor can you pair with strawberry to make a new combination?

   chocolate

2. Does pairing a scoop of strawberry with a scoop of vanilla make a new combination? Why or why not?

   no, because strawberry and vanilla is the same as vanilla and strawberry

3. How many different combinations of two different flavors of yogurt can you make with three flavors?

   3 combinations

4. Assume banana is added as another choice. Fill in the table to list all the combinations that can be made using two different flavors.

   Answers may be listed in a different order.

Scoop 1	Scoop 2	Scoop 1	Scoop 2
vanilla	chocolate	chocolate	strawberry
vanilla	strawberry	chocolate	banana
vanilla	banana	strawberry	banana

49

Puzzles, Twisters & Teasers
Banana-Rama!

Find the number of possible combination of pairs for each set of items below. Match the letters with the correct answers to solve the riddle.

1. peas, potatoes, corn, carrots, beans

   combinations: 10 A

2. Call of the Wild, Treasure Island, Peter Pan

   combinations: 3 E

3. bananas, apples

   combinations: 1 N

4. N'Sync, Britney Spears, Backstreet Boys, Janet Jackson

   combinations: 6 X

5. chocolate, vanilla, strawberry, pineapple, butter pecan, toffee, mint, blueberry, raspberry, cherry

   combinations: 45 B

6. Rob, Karen, Sari, George, Lemuel, Jenny, Tom, Lesley

   combinations: 28 C

7. blue shirt, black shirt, yellow shirt, white shirt, green shirt, brown shirt, gray shirt

   combinations: 21 H

8. Winston Churchill

   combinations: 0 P

9. Monet, Manet, Renoir, Pissarro, Degas, Cassatt, Picasso, Van Gogh, Gauguin

   combinations: 36 T

10. tennis ball, golf ball, football, soccer ball, basketball, snow ball

    combinations: 15 S

How is a flamingo like a bunch of bananas?

They're both pink, E X C E P T
                    3 6 28   0

T H E B A N A N A S
36 21   45   1   10   15 .

50

---

1. In how many ways can you arrange the letters in the word NOW? List the permutations.

6 ways; now, nwo, own, onw, won, wno

2. In how many ways can you arrange the numbers 4, 5, 6, and 7 to make a three-digit number? List the permutations.

24 ways; 4,567; 4,576; 4,657; 4,675; 4,756; 4,765; 5,467; 5,476;

5,647; 5,674; 5,746; 5,764; 6,457; 6,475; 6,547; 6,574; 6,745;

6,754; 7,456; 7,465; 7,546; 7,564; 7,645; 7,654

3. Find the number of permutations of the letters in the word FOUR.   24

4. In how many ways can you arrange the numbers 3, 4, 5, 6, and 7 to make a five-digit number?   120 ways

5. Find the number of ways you can arrange the letters in the word *arrange*.   5,040

Choose the letter for the best answer.

6. What is another way of showing 5 factorial or 5!?
A 5^3
B $5 + 4 + 3 + 2 + 1$
Ⓒ $5 \cdot 4 \cdot 3 \cdot 2 \cdot 1$
D 5^5

7. How many permutations of the numbers 10 through 14 are there?
F 10!
Ⓖ 5!
H 4!
J 14!

8. In how many ways can 5 children be matched with 5 puppies?
A 5
B 25
C 100
Ⓓ 120

9. Six friends are waiting in line at the movie theater. In how many different orders can they be standing in line?
Ⓕ 720
G 120
H 36
J 6

10. How many permutations of the numbers 1 through 9 are there?
A 9
B 900,000
Ⓒ 9!
D 9^9

11. In how many different ways can seven drivers be matched up with seven rental cars?
Ⓕ 5,040
G 2,401
H 49
J 7

51
Holt Mathematics

---

1. Joe has homework assignments for math, Spanish, and history. In how many different orders can he do his homework?   6 orders

2. Find the number of permutations of the letters in the word SMART.   120

3. In how many ways can you arrange the numbers 6, 7, 8, and 9 to make a four-digit number?   24 ways

4. A table has 8 seats. In how many different ways can 8 people sit at the table?   40,320 ways

5. Nine mountain bikers are on a bicycle trip. In how many possible ways can they follow each other?   362,880 ways

6. Seven students are waiting in line at the cafeteria. In how many different orders can they be standing in line?   5,040 orders

7. How many permutations of the letters A through F are there?   720

8. Ed, Martine, Sal, Carl, Paula, Terry, Ken, Leo, Ursula, and Jamie are in a race. In how many different orders can they finish?   3,628,800

9. Find the number of permutations of the letters in the word *permutations*.   479,001,600

10. In how many different orders can 11 people stand in line?   39,916,800

11. In how many different ways can a librarian arrange eight books on a shelf?   40,320 ways

12. Melinda has 15 art trophies. Write an expression that shows how many different ways she can line up her trophies on a shelf.   15!

52
Holt Mathematics

---

1. a. Find the number of permutations of the letters in the word BRAIN.   120

b. If you choose one of the permutations at random, what is the probability that it will start with a vowel? $\frac{2}{5}$

2. a. In how many ways can you arrange the numbers 4, 5, 6, 7, 8, and 9 to make a six-digit number?   720 ways

b. If you choose one of the six-digit numbers at random, what is the probability that the number will be less than 600,000? $\frac{1}{3}$

3. Eight skydivers are on an airplane. In how many possible ways can they jump from the plane?   40,320 ways

4. There are seven different flavors of yogurt at a supermarket. In how many different orders can they be lined up on the shelf?   5,040 orders

5. How many permutations of the numbers 1 through 9 are there?   362,880

6. In how many different ways can a video store arrange ten videos on a shelf?   3,628,800 ways

7. Harrison has a collection of 25 antique teapots. Write an expression that shows in how many different ways he can display his teapots.   25!

Determine whether each problem involves permutations or combinations.

8. Elect four people to be president, vice president, secretary, and treasurer.   permutations

9. Order a three-topping pizza from a choice of eight toppings.   combinations

10. Decide how many ways you can put ten books on a shelf.   permutations

11. Organize 30 students into two-person doubles teams.   combinations

53
Holt Mathematics

A **permutation** is a selection of objects in a particular order.

In how many ways can Allie, Bob, and Carl stand in a line?

You can draw a tree diagram to find the number of permutations.

1. Complete the tree diagram.

Allie — Bob — Carl
Allie — Carl — Bob
Bob — Allie — Carl
Bob — Carl — Allie
Carl — Allie — Bob
Carl — Bob — Allie

2. Complete the list of the outcomes.

Allie Bob Carl   Bob Allie Carl   Carl Allie Bob
Allie Carl Bob   Bob Carl Allie   Carl Bob Allie

You can also use multiplication to find the number of permutations.

3. Complete the multiplication.

$$3 \times 2 \times 1 = 6$$

Choices for first in line   Choices for second in line   Choices for third in line   Number of permutations

4. There are   6   permutations for 3 people standing in line.

You can find the probability that they will be standing in line alphabetically.

5. $P(\text{alphabetical order}) = \dfrac{\text{number of alphabetical arrangements}}{\text{total number of permutations}} = \dfrac{1}{6}$

6. Find the number of ways you can arrange the letters in the word MATH.

$$4 \cdot 3 \cdot 2 \cdot 1 = 24 \text{ ways}$$

54
Holt Mathematics

Challenge
Factorial Fun

Sometimes you can use the definition of *factorial* to simplify computations with factorials.

Example: $\dfrac{8!}{5!2!} = \dfrac{8 \cdot 7 \cdot \overset{3}{\cancel{6}} \cdot \cancel{5 \cdot 4 \cdot 3 \cdot 2 \cdot 1}}{(\cancel{5 \cdot 4 \cdot 3 \cdot 2 \cdot 1})(2 \cdot 1)} = \dfrac{8 \cdot 7 \cdot 3}{1} = 168$

Simplify. Remember to use the order of operations.

1. $(3 + 2)!$ 2. $(4!)(2!)$ 3. $(2 \cdot 3)!$ 4. $5 + 4!$

   120   48   720   29

5. $\dfrac{6!}{7!}$ 6. $\dfrac{6!}{4!}$ 7. $\dfrac{5!}{7!}$ 8. $\dfrac{9!}{7!}$

   $\dfrac{1}{7}$ 30 $\dfrac{1}{42}$ 72

9. $\dfrac{8!}{3!5!}$ 10. $\dfrac{6!}{2!4!}$ 11. $\dfrac{9!}{8!3!}$ 12. $\dfrac{10!}{8!4!}$

   56   15 $\dfrac{3}{2}$ $\dfrac{15}{4}$

13. $\dfrac{8!}{4!4!}$ 14. $\dfrac{3!5!}{4!}$ 15. $\dfrac{11!}{7!4!}$ 16. $\dfrac{6!8!}{7!5!}$

   70   30   330   48

17. $\dfrac{2! + 2!}{3!}$ 18. $\dfrac{3!3! + 3}{3!}$ 19. $\dfrac{4!}{3! + 4}$ 20. $3\left(\dfrac{3!}{4!}\right)$

   $\dfrac{2}{3}$ $\dfrac{13}{2}$ $\dfrac{12}{5}$ $\dfrac{3}{4}$

21. $2!(3!)$ 22. $4! - 3!$ 23. $4!(3!)(2!)$ 24. $5 + 4! + 3$

   12   18   288   32

   55   **Holt Mathematics**

Problem Solving
Permutations

Write the correct answer.

1. Five snowboarders are competing in a half-pipe competition during the Winter Sports Festival. In how many different orders can the snowboarders compete?

   120 orders

2. In how many different orders can, Debbie, Brigitte, and Adam wait in line in the school cafeteria? What is the probability that they will be in alphabetical order?

   6 orders; $\dfrac{1}{6}$

3. In how many different orders can the science class study the planets Jupiter, Saturn, Uranus, and Neptune? What is the probability that they will study Saturn first?

   24 orders; $\dfrac{1}{4}$

4. The physical education teacher sets up 8 different exercise stations for the class to complete. In how many different orders can the stations be done?

   40,320 orders

Choose the letter for the best answer.

5. Hannah, Javier, and Beth were the three qualifiers for a race. In how many different orders can they finish? What is the probability that Hannah or Beth will be first?

   A 3 orders; $\dfrac{1}{3}$

   B 3 orders; $\dfrac{2}{3}$

   C 6 orders; $\dfrac{1}{3}$

   Ⓓ 6 orders; $\dfrac{2}{3}$

6. Berto and 5 friends have ordered dinner at a restaurant. What is the probability that Berto will be served last?

   F $\dfrac{1}{720}$

   G $\dfrac{1}{120}$

   H $\dfrac{1}{60}$

   Ⓙ $\dfrac{1}{6}$

7. Six different prizes are being awarded to the winners of a contest. In how many different ways can the prizes be awarded?

   A 4,320 ways   C 120 ways

   Ⓑ 720 ways   D 30 ways

8. Jonathan will have math, English, history, social studies, and science each day next year. What is the probability that his classes will be in alphabetical order?

   F $\dfrac{1}{5}$ Ⓗ $\dfrac{1}{120}$

   G $\dfrac{1}{24}$ J $\dfrac{1}{720}$

   56   **Holt Mathematics**

Reading Strategies
Make an Organized List

An arrangement of objects in a certain order is called a **permutation**.

How many different ways are there to arrange turkey, lettuce, and tomato on a sandwich?

You can make a list of the items just as they would appear on the sandwich. If turkey is on top, the lettuce and tomato could be arranged as follows:

Turkey   Turkey
↓   ↓
Lettuce Tomato
↓   ↓
Tomato   Lettuce

Answer each question.

1. List the ways you can arrange the three items with lettuce on the top.

   lettuce, tomato, turkey; lettuce, turkey, tomato

2. How many different arrangements are there with lettuce on top?

   2

3. List the ways you can arrange the three items with the tomato on top.

   tomato, lettuce, turkey; tomato, turkey, lettuce

4. Make an organized list showing all the different arrangements that can be made with turkey, lettuce, and tomato.

   turkey, lettuce, tomato   turkey, tomato, lettuce

   lettuce, tomato, turkey   lettuce, turkey, tomato

   tomato, turkey, lettuce   tomato, lettuce, turkey

5. How many different ways can you arrange tomato, lettuce and turkey on a sandwich?

   6 different ways

   57   **Holt Mathematics**

Puzzles, Twisters & Teasers
Stay on Track!

(Crossword puzzle answers)

Across: 1. PERMUTATION   5. CHOICE   7. FUNDAMENTAL   8. COMBINATIONS

Down: 1. POSSIBLE   2. RANDOM   3. ALPHABETICAL   4. LIST   6. FACTORIAL

Across

1. A _____ is an arrangement of objects or events in which the order is important.

5. In a permutation, each time you make a _____ for one position, there is one less from which to choose.

7. The _____ Counting Principle can help you find the number of permutations.

8. Permutations and _____ are not the same.

Down

1. The number of _____ permutations of a group of objects may be written as a factorial.

2. The order of a group of objects may be chosen at _____.

3. An _____ permutation is based on the letters in the names of the objects.

4. You can use a _____ to find the number of permutations of a group of objects.

6. You can find the _____ of 6 by multiplying $6 \times 5 \times 4 \times 3 \times 2 \times 1$.

   58   **Holt Mathematics**

---

   72   **Holt Mathematics**